THE HIGHLAND

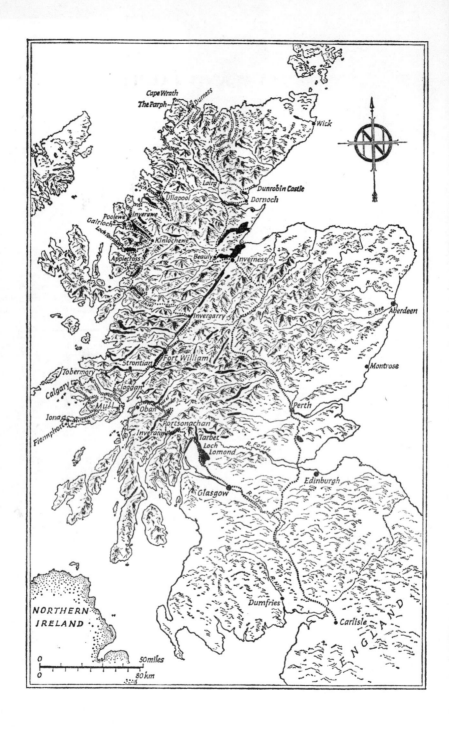

CapeWrath
The Parph
Durness
Wick
Lairg
Ullapool
Dunrobin Castle
Dornoch
Poolewe
Inverewe
Gairloch
Loch
Kinlochewe
Applecross
Beauly
Inverness
R. Spey
R. Don
R. Dee
Aberdeen
Invergarry
Montrose
Strontian
Fort William
Tobermory
Calgary
Claggan
Mull
Oban
Perth
Iona
Fionnphort
Inveraray
Portsonachan
Tarbet
Loch
Lomond
R. Forth
Edinburgh
Loch Fyne
Glasgow
R. Clyde
NORTHERN
IRELAND
R. Nith
Dumfries
ENGLAND
Carlisle

0   50miles
0   80km

The
Highland Jaunt

PAUL JOHNSON AND
GEORGE GALE

COLLINS
ST JAMES'S PLACE, LONDON
1973

William Collins Sons & Co Ltd
London · Glasgow · Sydney · Auckland
Toronto · Johannesburg

First Published 1973
ISBN 0 00 211333 3

Set in Monotype Scotch Roman
Made and Printed in Great Britain by
William Collins Sons & Co Ltd Glasgow

*The drawings for the endpapers and
on pages 42, 131 and 134 are by George Gale;
the map by H. A. Shellez; and the rest
by Paul Johnson.*

PART ONE

Paul Johnson's Narrative

CHAPTER ONE

A favourable start – My companion – The Scotch mean
what they say – Time's winged motor-car – A
marina on Loch Lomond – The beginning of the High-
lands – Inveraray and the Argylls – The importance
of hypocrisy – Anxieties about the licensing laws –
Perils of Highland lettuce-growing.

George Gale and I had often talked of going on what Dr
Johnson called 'a Highland jaunt', and recording it in words
and pictures. But, just as often, the project had to be post-
poned, and when we at last actually found ourselves sitting in
George's car, heading for the North, our relief and satisfaction
was nearly as great as James Boswell's, when he first greeted
Dr Johnson on Scottish soil. It was early in June, 1971, when
we left London. The sun was shining, as it continued to do
almost throughout our expedition. We had a good, fast car,
which gave us no anxiety. That night, we slept at a comfortable hotel, on the shores of Ullswater, well placed for a
dash across the frontier the next morning. The omens seemed
favourable. We had driven north behind a fast and luxurious
hearse, transporting the mortal remains of a patriotic Scotsman to their Highland resting-place. Moreover, at breakfast
that first morning by the lake, we found, in an early edition of
The Times, a notable misprint in the paper's account of Derby
Day: 'The Queen, escorted by the Duke of Norfolk, inspected
the hores with a knowledgeable stare.' This we took as a sign
of joyful times to come, for we intended to inspect the Highlands, if not with a knowledgable, at least with an inquiring
stare.

Before I begin my account of our journey, the reader may
wish to know a little more about my companion, so I will
follow Boswell's example and 'attempt a sketch of him'.
George is now, like myself, in his early forties. He is a little

9

over the middle height, perhaps a little overweight, too, and certainly disinclined to vigorous physical exercise. But he is otherwise alert, energetic and most pertinacious. A ripe scholar of Cambridge, he has spent his adult life in the pursuit of information for the public, working for the *Guardian*, the *Express* and the *Mirror*. He now edits the *Spectator*, our oldest political weekly. His hair is dark red, worn long, and streaked with a becoming grey. His voice, to which he is attached, makes a noise like gravel being poured into a concrete-mixer. Because of his abrupt manner, and the fact that he does not suffer fools gladly, he is generally supposed to be rude, but he is in fact capable of unusual courtesy. Like Dr Johnson, he is 'correct, nay stern in his taste; hard to please and easily offended; impetuous and irritable in his temper, but of a most humane and benevolent heart'. He has travelled all over the world, and witnessed many tragic and bloody events. He has been placed in front of a Congolese firing-squad, been ill-treated by Irish soldiers for a supposed aspersion on the Virgin Mary, and declared *persona non grata* by various dictators. He has a peculiar hatred of the telephone, having suffered from its vagaries in many disagreeable places. As Churchill wrote of F. E. Smith, he has 'reached settled and somewhat sombre conclusions upon a large number of questions, about which many people are content to remain in placid suspense'. George distrusts idealism and grandiose theory; he believes, from experience, that most human beings are foolish, and many of them wicked; he has well-argued doubts about all schemes for the perfectibility of mankind and the inauguration of universal peace and happiness. Progress towards civilisation, he feels, is hardly won and easily lost. An agnostic by conviction, and a republican by instinct, he is also a Presbyterian by upbringing, and inclined to think that conservatives have a firmer grasp on realities than socialists. He dislikes pretentiousness, snobbery, inaccuracy, cant, loose thinking, do-gooders, humanitarians, readers of progressive newspapers, virtually all males between the ages of 17 and 25, and high-minded public figures such as Anthony Wedgwood Benn. He makes the most extraordinary grunting noises. He is also a fluent and incisive writer, and drives cars with speed and safety.

As for myself, I would not claim, like Boswell, to be 'completely a citizen of the world'. But I have seen most of it, and believe it can be changed for the better. I am a Catholic and a socialist, and look forward to the coming of Christ's kingdom on earth, an event which George thinks is most improbable. My heart and head are on the Left (I am left-handed, too), and I yield to none in my dislike for the capitalist system, Heathian suburbia, the public schools, the landed aristocracy, conspicuous consumption and the shooting of lame ducks. Like George, I have the disadvantage of possessing red hair, and therefore a wholly unfounded reputation for irascibility. George and I also share a profound respect for British sovereignty, which makes us both confirmed opponents of the Common Market. I do not drive cars, but I read maps skilfully, a claim George would dispute. Altogether, we were a well-matched pair, and did not once quarrel throughout our jaunt.

There is nothing much to crossing the Scotch frontier. You go over the bridge, and that is that. But we were clearly entering a new country. For one thing, the English do not usually say what they mean, especially in their road signs, which are the worst in the world, next to Japan. On the English side of this bridge there is a sign which reads: 'Dual Carriageway Ahead: Single Lane Traffic Only.' I do not know what this notice means; but it clearly cannot mean what it says. On the other hand, the Scotch are explicit. On their side of the bridge a sign says: 'Scotland's First House', which it undoubtedly is. It also says: 'Over 10,000 marriages performed'; and I do not question this either. When, shortly, we passed a building called 'Alexandra Parade Public School', we knew that this was a school actually open to the public and not, as in England, reserved for the children of the rich. And then the Scotch do not refer to 'Police Station', which sounds safe and cosy, and is misleading; they call it 'Police Headquarters', which sounds sinister, and is accurate.

Nevertheless, Scotch phraseology, being explicit, gives the game away. Anyone who reads through the glossary, in the admirable *Shell Guide to Scotland*, of 'Gaelic and other local words' – words chosen, presumably, because of their import-

11

ance and frequency – must get an impression of violence and
lawlessness. There is 'colleague', a good and flavoursome word,
but in Scotland a verb meaning 'to conspire'. There is a special
word for a fortified gate, and no less than three for a fortified
house. 'Cateran' is a freebooter, and 'reiver' a ravager or
plunderer. Even 'chaumer' means a police-court as well as a
chamber. Then there is 'cathair', a fortified position, and
'joug', described as an 'iron collar for prisoners'. 'Loon' means
a lad, and not a lunatic. More disturbingly, 'mortsafe' means
'a cast-iron frame to protect graves from body-snatchers'.
Now there can't be many countries where it has been found
necessary to invent a special thing to stop people from robbing
graves, still less to coin a word for it. Do Scotch undertakers
throw in a mortsafe as part of a package deal? These glossaries,
of course, have a taste for the bizarre. What exactly is a roof-
ridge, in Scotch 'riggin'? Is there much demand for 'leisters', or
barbed fishing-spears? If I asked for a 'cran', would I in fact
get a measure of herrings, or just a knowledgeable stare? How
often do the Scotch use a 'papingo', here defined as 'a wooden
parrot used as target for arrows'? Have they ever used many
papingos? Then there is the case of the 'gumely jaup', a
striking phrase which conjures up many possible meanings but
actually signifies 'a muddy spray'. Do these muddy sprays
occur frequently in Scotland? Are they directed at passing
English motorists by loons? I have now almost exhausted the
glossary, but am not much nearer a meaningful sentence.
'Reiver! Cease to colleague with your fellow-caterans to squirt
that gumely jaup on my father's mortsafe, or I will take you
to the chaumer and have you put in a joug.' It is the best I
can do.

We drove rapidly across the southern uplands, a land of small,
stunted hedges, many plantations, of evident poverty, though
the contrast with England is much less striking than in the
past. There were few people about: this is the territory of the
rich border lairds, who send their sons to Eton and sit in
Cabinets. In no time at all, the motorways to and beyond
Glasgow will be finished, and then English families will be able
to drive easily from London to the Highlands in less than a

day, and the remote places will be shattered for ever. We are on the very eve of a great transformation of the hills, which will soon lie at the feet of motorised millions from the vast conurbations of the English south and midlands. I reflected that George and I were coming just in time. Boswell took Johnson to the Highlands and Hebrides in 1773 'that we might there contemplate a system of life almost totally different from what we had been accustomed to see; and to find simplicity and wildness, and all the circumstances of remote time and place'. In fact they arrived too late: the Highland system was slowly decomposing; much of it had already gone for good by the 1770s. We, too, have been racing against the progressive clock, a little more successfully. By the time this account appears, certainly in a few years, the Highlands will be a different, and less pleasing, part of Britain. Time's winged chariot is hurrying near, powered by the latest mark of internal combustion engine.

Glasgow, indeed, once a powerful obstacle to motorised progress, is being torn apart: we drove through most of it on a great concrete bridge, spanning demolished sandstone slums. This grim and beautiful city, which George rightly calls the most foreign in all Britain, was nobly conceived in the first fine flush of the industrial revolution, a vast classical artefact carved in stern local rock; now they are nailing upon it a high superstructure of fast roads, as in any large American town of the Middle West. It will soon be nothing more than a visual incident – a flash of urban scenery – on a rapid thrust to the hills and lochs. Around lies a glittering panorama of skyscraper flats, into which the dispossessed poor of central Glasgow are being decanted. Beyond it was an immense and graceful bridge, which I sketched, not yet open when we passed but now another link in the chain which is civilising, or incarcerating, the wilds: a *joug*, in short.

The alarming thing about Glasgow is that it is on the very verge of the Highlands. A few minutes brought us to the shores of Loch Lomond. There, Boswell relates, 'After breakfast, Dr Johnson and I were furnished with a boat, and sailed about upon Loch Lomond, and landed on some of the islands which are interspersed. He was much pleased with the scene.'

And well he might be. It is still a pleasing scene, and there is
no through road on the far side of the loch, which sparkled
under a blazing sun. But the affluent society has already
lapped its shores. Myriads of little, brightly-coloured sailing
boats bounced on the water; speed-boats roared to and fro;
and we called at the Duck Bay Marina, from which such
activities radiate. There is a vast bar and restaurant, whose
plate-glass, glare-proof windows frame the water and the hills
beyond. Teams of smart and pretty waitresses, in tartan
mini-kilts, busied themselves serving scampi and chips, and
other traditional Scotch dishes. There were thousands of
people about, and hundreds of cars. A shop sold tartan every-
things, and seven-year-old whisky marmalade. But there was,
and I am afraid to say I was glad to see it, a residual hint of
the Kirk, for a notice on the Marina stated firmly: 'No persons
under 21 admitted after 7 p.m.' If you are eighteen in Scotland
you may vote for the Union parliament, and fight in the
English army, and go to prison; but you can't get into the
Duck Bay Marina after the children's hour.

It is a matter of argument whether Loch Lomond is in the
Highlands. Some would say that it is beginning to look like
Glasgow's boating lake. But there can be no dispute that when
you cross Glen Croe and descend to Loch Fyne, and reach the
little town of Inveraray which lies on its shore, you are in
Highland country. The English do not realise that Scotland is
two countries and two races, and that the difference is not
simply a matter of altitude but of history. Beyond the Highland line, which runs from south-west to north-east, the people
are Celts, with a strong dash of Norse blood. Their culture is,
or was, Gaelic, their social structure tribal and patriarchal The
truth is that in the eleventh and twelfth centuries, the Scottish south-east – including the lowlands and the southern uplands – underwent a Norman conquest which in some ways was
more profound and permanent than anything experienced in
England. Norman knights and dispossessed English thegns,
Flemings, Bretons and men from central France, moved in
from the south and established a territorial system of military

government which bore a much closer resemblance to the academic abstraction called feudalism than the forms which William the Conqueror imposed on the English. In England, the Normans inherited an ancient centralised monarchy and administration, which successfully survived feudal devolution. In south-east Scotland, the coming of the Continentalised culture brought about a revolution. The Church was transformed on Hildebrandine lines, and resisting Gaelic clerics driven out. A personal monarchy was imposed, all land declared possessed of the Crown, and military tenures carved out of it in so thorough a fashion that even to-day Scotch land law has a residual feudal basis: people who would be called freeholders in England pay 'feu duties' in Scotland – and very irksome they are.

Above all, the newcomers built towns, or burghs, settled by a mixed breed whose common language was an English dialect. English became the language of commerce, and of government, and ultimately of learning, and in the process the universal tongue of the inhabitants. The Gaelic race and the Gaelic culture were superseded long before the end of the Middle Ages. In 1380 the Aberdeen chronicler, John of Fordun, divided the nation into 'the wild Scots and the householding Scots'. But some lowlanders would not allow that the Highlanders were Scots at all: they themselves were 'the Scots'; those who lived beyond the Highland line were 'the Irish' – and it is true that the Scotch gaels came originally from Ireland. The shape of modern Scotch society was created by the lowlanders on an English basis, against which the Highlanders formed an ultimately futile force of resistance. The dynamic of the Scotch reformation was the English-speaking burghs. It was the burghers and their lowland land-owning allies who threw off the French and Continental connection, and forged the alliance with England which eventually produced the Act of Union of 1707. Modern Scotland is a creation of English culture, whether the Scotch like it or not.

On the other hand, north-west of the Highland line, the Anglo-Norman invasion could not penetrate. The people remained Gaelic in race and speech. A few of the Norman barons transformed themselves into Highland chieftains, and accepted the social structure of the country. In the later Middle Ages, the Lords of the Isles were virtually a second royal dynasty, and held their court until the last of them was destroyed by James IV in 1493. The Highlands and Islands were slowly absorbed into the Scottish state, and then in the eighteenth century into the British state. But the people have retained a separate identity; they look back on a different history; and they cling to the relics of a culture which is in many ways alien to the rest of Scotland.

Now in this historic battle of cultures, Inveraray played a key role. It was, and is, a very special town – and it looks it. In the eighteenth century it was a frontier post: the most northerly town, indeed the only town, on the whole western coast. It was the prime instrument of penetration for English political and social civilisation, the miniature capital of the great Campbell clan of the Dukes of Argyll, who triumphantly upheld for more than two centuries the interests of the Protestant south. There is no finer sight in the British Isles than the moment when you turn the bend of the hill and see spread before you, on gently sloping and wooded ground, the huge toy castle of the dukes, and the eighteenth century silhouette of the town at its feet. In size it is no more than an English village. In conception, in dignity, in purity of line and unity of execution, it is the most splendid town in the kingdom, entirely worthy of its historical importance.

When Boswell and Johnson reached Inveraray, after their wanderings in the islands, they felt they had regained civilisation. There, said Johnson, 'we found an inn, not only commodious, but magnificent.' 'An excellent inn,' echoed Boswell: 'we supped well.' This inn remains; it is now the Argyll Arms, and it was there that the abstemious Johnson, for the first and last time, called for a gill of whisky with the immortal phrase: 'Come, let me know what it is that makes a Scotchman happy!' George and I, too, were happy in Inveraray. The sun was hot; the sky was cloudless. We looked at the noble line of the town, designed by William and John Adam, and completed by Robert Mylne, and pronounced it perfect.

There is a remarkable church in Inveraray, built by Mylne

in 1794, and evidence both of fine taste and of the spirit of
toleration which the Dukes of Argyll brought to the Highlands.
It is a double-church, that is to say two churches, exactly
alike, partitioned by a dividing wall. On one side the Presbyterian services were conducted in English, according to the
lowland form; on the other in Gaelic. The two congregations
had come into separate existence in 1650, and remained so
until 1930, when, Gaelic having virtually died out there, they
were united; and now the Gaelic side is unused, indeed the door
was locked. The 'English' church is impressive: a great,
echoing auditorium, in which the altar is nothing, and the
pulpit everything. I do not know whether the Reverend John
McCaulay, one of the ministers at Inveraray in 1773 – presumably the English-speaking one – lived to rant in the new
church. Perhaps he was not a ranter. He submitted cheerfully
to a magisterial rebuke from Dr Johnson ('Sir, are you so
grossly ignorant of human nature, as not to know that a man
may be very sincere in good principles, without having good
practice?'), and returned to breakfast the next morning,
'nothing hurt or dismayed by his last night's correction'. I
have a feeling that ministers at Inveraray were not encouraged
to take too much upon themselves.

At the entrance to the church are two wooden staircases,
one labelled 'The Duke's Loft', and the other 'The Magistrates'
Loft'. I went up into the Duke's, and George into the Magistrates', and in the Duke's I found four high and stately chairs,
upholstered in red; the magistrates sat on more humble stools.
From here we looked down on the pulpit; and from below it
would take a strong-minded clericalist to offer defiance to the
assembled dignitaries of the state glaring down from above.
The eighteenth century set-up in Inveraray, with the Duke
perched in his castle, or on Sundays in his loft, emphasised the
salient characteristic of the English connection: the supremacy of the secular arm. The English do not like theocracy:
for them, religion is a matter not of piety, but of decorum, a
force for social order and outward observance. The Dukes of
Argyll were in this tradition, and so, of course, was Dr
Johnson. In rebuking Mr McCaulay he made an important
point: that hypocrisy is socially and morally preferable to open

and unashamed vice. This was the prevailing view of sensible men in the eighteenth century; it was destroyed by Charles Dickens, and has remained unfashionable ever since. But I think the doctor was right; and so does George.

There is a fine classical public building in Inveraray, once the courthouse where the Duke enforced the English connection, now in lamentable decay; the Highland Industries Board hope to restore it as a centre for training young people. I suppose the Dukes are no longer rich enough to keep the town in perfect repair. But nearby, the Ministry of Works, and a gifted Scotch architect, Mr Ian Lindsay, have carried out an admirable piece of restoration on a block of tenements, which have been made habitable, even luxurious, for ordinary people, while retaining the façade intact. It must have cost a fortune, and was worth it. This block has a common garden, where the washing hangs, and children play in safety; it made a domestic scene of eighteenth-century plebeian contentment, to rejoice the heart of a right-thinking Whig Duke; but, of course, made possible by democratic taxpayers and the welfare state. This block, however, demolished my theory that the Scotch mean what they say. For the tenements are called 'lands'. If you call houses lands, why not lands houses? And why do they call gardens 'policies'? What is political about a garden?

George and I were delighted with the Duke's castle, which is open to the public and is a little miracle of eclectic architecture. I am not sure what one should call it. You might say it is the earliest example of neo-Gothic in Britain. But, equally, it is a forerunner of Scotch baronial. It is big enough to be grand, and small enough to be comprehensible, and cosy. Its presence on the shores of a wild Highland sea-loch gives one faith in the power of civilisation, or at any rate of ducal money. The man who built it, Duke Archibald, was, according to Johnson, a mean man; that is, he was 'narrow' in what the Doctor called 'his quotidian expenses'. But when Johnson actually inspected the castle he exclaimed: 'What I admire here, is the total defiance of expense.' This was my feeling, too. It seems to me wholly wrong that people should be very rich to-day, especially since the houses they build are rubbish.

But I am glad they were very rich in the eighteenth century. Of course Boswell's eyes were elsewhere: 'I never shall forget the impression made upon my fancy by some of the ladies' maids tripping about in neat morning dresses.' And he adds: 'I could have been a knight-errant for them.' I shudder to think what effect the mini-kilt, or tartan hot-pants, would have had on Boswell. But I saw no mini-kilts tripping about the castle. The thing there is weapons: there must be hundreds, perhaps thousands, of them hanging on the walls, enough to polish off at a stroke (as Mr Heath would say) the entire population of Argyll. George remarked that, for a castle which has had its share, perhaps more than its share, of domestic disagreement, the ubiquitous presence of this formidable array must constitute a standing temptation.

From Inveraray we drove north into the wilds, and for the first time George began to display anxiety about the Scotch attitude to alcohol. I would not describe George as a drinking man; but he is a man who likes drink to be available. It gives him a proper sense of security to feel that, wherever he is, drink can be summoned at immediate notice, and without qualification. Now the Highlands are an area where this sense of security is notably lacking. Neither I nor George know the intricacies of the Highland licensing laws; not many people do, I imagine. But we were both aware that a great deal of human and Kirk-like ingenuity has gone into separating Highland man, and so Highland tourist, from drink on every possible occasion, and we feared, and specially George feared, to become the victims of this activity. That night we planned to sleep at a solitary hotel on a remote loch. I knew that the hotel existed and that rooms were available for us. But I had lost, or George had lost, the book in which the hotel was listed, and I could not faithfully swear that it was fully licensed. In England one would take such a thing for granted; but the Highlands are full of legal pitfalls for unwary travellers. Moreover, to judge from the map, this hotel was the one inhabited place for miles around, and once there we would be stuck.

George's fears were very much increased when, by ill-

chance, he seized my copy of the *Good Food Guide* and began
to study it. George is not a great believer in the *Guide*; he says
it encourages ignorant restaurateurs to put unsuccessful
French messes on otherwise wholesome English food. I take
the more optimistic view that it prevents a situation which is
notoriously bad from becoming still worse. However that may
be, George studied the *Guide* and found our hotel was not in
it, which he perversely interpreted as a bad sign. Moreover, his
eye alighted on some completely different place in Skye where,
he read out, 'the Misses Nicolson' were 'conducting a fine
rearguard action on behalf of the pudding'. George is not a
pudding man, and he greeted this information with some
derision; he dwelt at length on where the Misses Nicolson
could put their puddings. And then, to his horror, he saw at
the end of this entry the sinister phrase: 'With 12 hours'
notice, wines will be ordered.' No mention of whisky, he noted,
or any other alcoholic beverage. Moreover, if the Misses
Nicolson could not guarantee to provide wine at less than
twelve hours' notice, might not other, more slothful, pro-
prietors require twenty-four hours, or even a week? Might not
some decline altogether, and on principle? Were there, perhaps, special laws, stipulating the length of time which must
elapse between the ordering of a drink and its provision –
delays which would grow greater the farther north we pro-
gressed? George's gloom was not lightened by the numerous
signs we passed which said: 'Temperance Hotel', which he
took to be a Highland tautology, and one of which indeed
added 'And a true Highland welcome'.

But all this worry was needless. The hotel proved to be old,
and comfortable and licensed to the hilt. There were many fine
and rare malt whiskies in the bar, together with some remarkable wooden carvings, and a convivial company. There
we drank *Glen Morangie,* and toasted Boswell and Johnson.
The hotel stands directly on the loch, on an old ferry-crossing,
no longer in use. It is as remote, and agreeable, a Scotch
fishing-inn as anyone could wish for. But in 1773 it barely
existed. The day on which Boswell and Johnson reached it was
tempestuous even by Highland standards, and inspired a
notable purple passage in the Doctor's account. Boswell

merely records: 'We crossed in a ferry-boat a pretty wide lake, and on the farther side of it, close by the shore, found a hut for our inn.' He adds, with feeling: 'We were much wet.' Johnson, with what Boswell calls 'a species of heroism', refused to take off his clothes, 'letting them steam before the smoky turf fire'. Not surprisingly, it was later that evening, at Inveraray, that he felt the need for whisky.

The hotel has a little garden, sloping down to the loch. The sun was still high in the heavens at ten o'clock in the evening, and there, the next morning, which was fresh and glittering, I made a sketch of the huge panorama of loch and mountains. At breakfast we ate true Scotch porridge, which looks nutty, and tastes nice, but, as George said, is 'hard work'. The hotel is so remote that it has to be, I suppose, as self-supporting as possible. At any rate, it makes valiant efforts to grow its own lettuces. The lettuces have attracted rabbits, of which there are vast numbers around. The hotel imported cats to keep down the rabbits. But the cats intermarried, and multiplied, and now there are sixty of them running wild. Huge dogs were procured to deal with the cats, and indeed still lumber heavily about the hotel; but they do not notably, or visibly, chase cats, and the balance of nature remains unsatisfactory. The result is that the hotel reckons it costs them ten shillings to grow a lettuce.

During the night, four stags broke into the kitchen garden and ate the lot.

CHAPTER TWO

*Castles and solitude – The fury of the islanders – De-
population in the Hebrides – The dolorous island of
Mull – Iona and the Celtic church – Religion and
Left-wing politics – The greatness of Thomas Telford
– Our sense of insulation aroused*

Our route now lay along the eastern shore of Loch Awe, which
is a fine, long and deep stretch of water, once the north-
western frontier of Campbell territory, and the scene of much
fighting in the fifteenth and sixteenth centuries. The Campbells felt safe behind this natural defence, but they built many
fortifications to reinforce it; and the ruins of several fine
castles can be seen. George and I are both castle-fanciers, but
our tastes vary. On the far side of the loch I caught sight of the
kind of house I would like to inhabit: a choice and massive
piece of Victorian baronial, in full working order, with lavish
pinnacles, turrets and machicolations. A castle indeed, but
internally equipped with all the apparatus of nineteenth-
century comfort: solid teak doors and staircases, copper-hung
kitchens and reliable plumbing, a castle built not for siege but
to entertain an ample house-party at the height of the
Balmorality epoch. How such establishments still function
without staff I do not know; but perhaps servants can be found
even to-day, for the Highlands remain, to some extent, an
area of servility.

But George called my ambition vulgar, as indeed it is. He
prefers the genuine article, and at a turn of the loch found one:
a majestic ruin with some towers and buttresses standing, but
plainly uninhabited. This I sketched, against its setting of
water, sky and mountain. The trouble with such a ruin is:
how can it be made habitable without *ruining* it? On the other
hand, what is the point of having such a castle if you can't
live there? At Dunvegan, Johnson and Boswell took a high

23

romantic line with Lady Macleod, and insisted that she should continue to live on the castellated rock from which her ancestors had dominated Skye. But she, practical woman, pointed out that there was no room for a garden, that it 'must always be a rude place'; and that it was 'a *Herculean* labour to make a dinner there'. She thought it unfair of them to 'think of chaining honest people to a rock'. Well, then, what is the solution? Near Oban, someone appears to have found one. There we saw a ruined castle on a promontory; and tucked away within it is a modern stone house, in which the owner lives in comfort, surrounded by his jagged walls and turrets.

The wild Highlands have always tempted those in search of solitude. Even to-day it is possible to achieve it with a frightening degree of completeness. Some of the lochs have no road at all by their shores. We heard of an English lady who lives alone miles along remote Loch Etive, linked to civilisation only by a track along which no vehicle can pass. There are hundreds of such places in the West, where ruined crofts and bothies can be restored, and their owners can rejoice in a privacy which the twentieth century has come to think of as unattainable. Many attempt such a life; few survive even one long, hard and dark winter without an uncontrollable urge to return. To attain even an elementary degree of comfort in these savage places is a terrible, time-consuming struggle; an expensive one, too. Before his tragic death, even such a resourceful and determined man as Gavin Maxwell had found the effort to maintain himself at the place he called Camusfearna almost beyond his powers. His books tell a sad tale, though they make superb reading. The truth is, most of us at some time feel a hermitical urge to escape; but it is more sensible to release it vicariously.

At Oban, a busy and cheerful town and the hub of many sea-lanes, we took ship aboard the car-ferry to Mull. These ferries are run by MacBrayne's, who have what amounts to a monopoly, and are a source of acrimonious debate among Highlanders and Islanders. MacBrayne's ferries are popularly supposed to be enormously profitable; the probability is that they barely survive without a loss, for the season is painfully short and they must maintain a service through the endless winter months. But the ferries are expensive. We paid nearly £5, without passenger tickets, to carry the car over a 45-minute journey. Moreover, the day we were there large increases had been announced. It was the one subject of discussion throughout western Scotland. The *Scottish Daily Express*, a much more rumbustious organ than its English sister, had only one word in its headline, set in two-inch type: 'FURY!' 'The Isles,' it said, 'hit out at MacBrayne's fares rise.' It quoted one provost as saying that he was 'sickened with the mendacity' of the Scottish Office and MacBrayne's, a fine Presbyterian phrase which would not come easily to the lips of an English mayor.

Transport is the chief, the overwhelming problem of the Islands. Without some form of subsidy it is impossible to maintain a regular service at all. But can the taxpayer be expected to provide ever-increasing sums for an ever-diminishing number of people? As sea-transport becomes more costly, more and more islanders flee to the mainland to escape the financial burden and the restrictions of the time-table. So the losses mount, and there is no escape from the vicious circle.

I believe the population of all the islands is falling. Ulva, for instance, once supported 600; in 1851 there were 204; now less than a score. The house where Dr Johnson slept is burnt down. In 1837 there were 168 people on the island of Gometra; now two. Pabbay has one inhabitant, where once 300 lived. The Monach Islands now have no people at all. Mingulay had 142 even as late as 1891; now only a handful. Many islands, where communities flourished, are deserted to-day, or occupied for brief spells by adventurous people who take up the challenge but find the practical difficulties too great. In the *Oban Times* I read a sad little obituary of a Mr MacDonald of Carna, who 'farmed the island alone for many years during which time he was the sole occupant'. It ended: 'He was a fine piper, and a great character.' But such brave and solitary souls are rare; all who have tried island life agree it is vital to have good neighbours in time of trouble. Costly and determined efforts have been made to halt the depopulation of the islands. Romantic millionaires have spent fortunes on establishing local industries; the State has built roads and schools. But the affluence and the variety the young now demand cannot be provided. Even the larger centres are declining, despite a vast and growing tourist influx. Tobermory, the capital of Mull, had 1,500 people in 1843, 900 in 1947, and less than 700 in 1969. It is still falling, and this despite the fact that huge sums have been spent on improving the island's roads which are now, indeed, far better than one might expect.

I do not find this drift away surprising. We travelled all over Mull, which is remarkable for its soft beauty, strange rocks and rolling moors, dominated by the great Ben More, which changes its shape and colours as you circumnavigate it. But the distances are enormous. In the south we drove for over forty miles and passed only three places which could, by any standards, be called places of conviviality. One was described by a local guide as famous for its gaiety and its 'unusual forms of entertainment', but we found none – and that on a Friday night, which I presume is the best time to go there.

The people of Mull certainly used to live long lives. At a tiny wayside cemetery we found a score or more tombstones of those who survived eighty, and several ninety, years. But their descendants show a certain lassitude. Few seem to bother to grow fresh fruit or vegetables. At a small hotel we ate tinned soup, re-heated frozen beef, frozen vegetables, instant mashed potatoes and tinned fruit. There was nothing to drink, and the nearest pub fifteen miles away.

Moreover, the atmosphere is in some ways oppressive. While we were sitting in a bar, two policemen entered, scrutinised the faces of all present narrowly, then left without a word of explanation. Their mission may have been wholly benevolent, but it left a chill in the room. The kirk, too, is omni-present. A local paper devoted over 4,000 words to a report on its General Assembly. There were lengthy and abstruse discussions on the dangers of episcopal intrusion, and other echoes of ancient controversies. Do the young people care a rap for such things, when they are denied facilities which elsewhere in Britain are taken for granted? Nor do the lairds help much. No doubt the richer ones run their estates at a loss; or so they claim. But most care more about the deer and the fish than the people. One on Mull has been accused of taking down right-of-way signs to discourage walkers. Another sought to conceal the existence of a Telford chapel, for fear that its publication in a guidebook would attract visitors. In Tacitus's words, appropriately inserted in the mouth of a Scotchman, 'They make a desert and they call it peace.'

But one's sympathy for the ordinary islanders wanes when one surveys their feeble efforts to stand up for their rights and dominate their environment. Those with a sturdy independence of mind have long since left. As Johnson remarked on Mull: 'The lairds, instead of improving their country, diminished their people.' But why did the people allow themselves to be diminished? To be sure, Johnson had a low opinion of Mull, judging places, as he did, by the number of trees there, and finding precious few on this lovely island. When he lost his walking stick, he rejected assurances that it would be found and returned to him: 'Consider, Sir, the value of such a *piece of timber* here!' Mull, as George and I saw it, radiated extraordinary beauty. The sea was calm and deep blue. The hills and lochs blazed under a Riviera sun. On the silver-white beaches, people were still bathing after supper, and the sunset was one of those prolonged pink-and-turquoise affairs best left undescribed. But we met a shepherd who was returning to his lonely croft with his four graceful dogs and a new bottle of whisky. What else was there to do? I can't help agreeing with

27

Johnson's verdict on Mull: 'O, Sir, a most dolorous country!'

On the western tip of Mull is the little ferry-head of Fionnphort, and a mile across the Hebridean sea – it looks much nearer – the island of Iona. We had slept the night at a sort of boarding-house, which we had not found congenial. The silence in the dining-room, where blue-and-yellow porcelain birds flew across the walls, was glacial; no noise except the click of knives and forks, and the efforts of fellow-guests to deaden the sounds of mastication; the food was tasteless. But the next morning, as always on this jaunt, was spectacularly fine. A tiny ferry plies between the islands, and in this we chugged across to Iona.

Dr Johnson has a famous passage on Iona which ends: 'That man is little to be envied, whose patriotism would not gain force upon the plain of *Marathon*, or whose piety would not grow warmer among the ruins of *Iona!*' His words greatly struck contemporaries. Boswell thought that they alone justified the Doctor's book; and they have been carved in stone upon the island. There is, indeed, something wonderful that, from this small and remote place in the sixth century, on the very rim of Europe, a great religious movement should have radiated all over the British Isles. Considering the distances, and the means of transport, and the poverty of the place, the achievement of St Colomba and his monks was prodigious: bringing to vast areas not just a complex faith, but civilisation itself – the art of writing and of reading, the use of Latin, verse and music, architecture and carving, illumination and mathematics.

Iona was killed, long before the Reformation, by the Church of Rome. The decentralised, non-episcopal church for which it stood, with its strong sense of local communities and its Gaelic culture, was anathema to the Roman clergy. Iona received its first setback at the Synod of Whitby, which Colomba's successor, Coleman, left in disgust when the Northumbrian authorities accepted the Roman rules. The Celtic church there survived for a time through its inaccessibility. But after lowland Scotland was Normanised, and the Hildebrandine reforms introduced, the expulsion of the heterodox monks became a prime object of the Roman hierarchy. In the thirteenth century the aim was achieved. Regular Benedictine monks and orthodox nuns took over, under cover of the secular sword, and the peculiar spirit of Iona was lost. The church there simply reflected the forms and assumptions of feudal society. So the Reformation, when orthodox Catholicism on Iona was destroyed, was a long-delayed revenge. In 1899 the Duke of Argyll gave the ruins to the Church of Scotland; and the community which has since developed there is closer to the original genius of the place than the Roman usurpation – for the Scotch kirk, with its distrust of bishops and hierarchies, has some of the independent localism of Colomba's monks.

All things considered, the great church has been wisely and faithfully restored. It is a good place in which to pray, for it conveys no sense of belonging to any particular communion, while avoiding the anaesthetised neutrality of ecumenical church-boxes. I asked a blessing on family and friends and on various Left-wing causes. In a side-chapel is a marble effigy of the Duke-donor, together with the framed deed of covenant in which he transferred the land, an elaborate and impressive example of Scottish juridical prose. The community is not, strictly speaking, contemplative, for its members perform social work in the Glasgow slums and elsewhere. It has a radical tinge, and campaigns against nuclear weapons and war-mongering. The atmosphere is earnest, benevolent, worthy and homespun – one that some nasty woman-reporter would take malicious pleasure in ridiculing. You can, indeed, join the community, and some young men working in the church had evidently done so. But the community also takes in guests in July and August, at the modest fee of £5.50 a week. There is an elaborate programme of meditation, group discussions, and so forth. But I fear that most of those who go there are old: a form asks you to state whether you are under 30, under 70, or over. In the cloister coffee-shop, some elderly people were exchanging ideas. One lady said she had planned to kill herself before coming to Iona, to escape from 'the lawyers' who were harassing her. The others had had legal difficulties, too. It was all to do with unpaid telephone bills, perhaps a common Scottish problem.

The connection between religious enthusiasm and radical politics puzzles most people, but seems to me natural. What more appropriate that the holy men who live on Iona should develop a peculiar detestation of H-bombs which kill millions in the futile pursuit of material power! Until recently, the Roman Catholic church was associated with the darkest reaction all over the earth. But now that Rome has lost its arrogant self-assurance, all the best young priests have won the freedom to place themselves on the Left. In the United States, they cheerfully go to prison on behalf of the Vietnamese people. They suffer torture in Brazil. Even in Ireland, under an obscurantist hierarchy, and a social and political system which hates and fears any change, more and more priests are joining the reform movements. Perhaps one day a Pope will get himself martyred in the cause of the oppressed, and start a genuine revival of Christianity; nothing else will do it. All the same, some curious anomalies arise when churchmen join the Left. In Israel, the religious parties, whose strict sabbatarianism makes the Scotch Presbyterians seem easy-going, work hand in glove with the radical wing of the Labour Party, which demands secularisation of the State – a cynical alliance. And in East Anglia, the really High Church vicars and ritualists have a traditional sympathy for communism. I once met an ancient and saintly rector there whose regard for Josef Stalin was exceeded only by his veneration for Charles the Martyr.

Restoring all the bits and pieces of ancient Iona presents difficulties which are often insuperable. We admired a fine tenth-century cross which is undoubtedly genuine, but in a morning spent poking around among the tombs and masonry, we found it difficult to separate cherished myth from historical fact. There is a simple stone causeway known as the Road of the Dead, a sepulchral chapel, and many burial-stones of great antiquity. Kings, chieftains and lairds are supposed to be interred here, a highly-regarded privilege in the Dark and Middle Ages. But nothing is certain in the confusion, which has been tidied up rather than elucidated. Johnson's cool verdict is probably the best one: 'The graves are very

numerous, and some of them undoubtedly contain the remains
of men, who did not expect to be so soon forgotten.' Here,
Boswell suffered disappointment, for he foolishly supposed
the monuments would be as splendid and elaborate as those in
Westminster Abbey. But it is a serene and dignified resting-
place; one could ask for no better. Nearby, the ruins of the
cloisters, once inhabited by unloved and unlovable nuns, have
been made into a magnificent flower-garden, which blazed
with colours. Wherever one looks in Iona there is nothing but
beauty, except for the occasional rusty corrugated-iron roof,
beloved of the islanders.

When Boswell and Johnson were here in 1773, there was no
kirk, and the Ionians were virtually pagan. Now there is a
simple and elegant church, with spacious Gothic windows of
plain glass, one of thirty-two which Thomas Telford scattered
throughout the Highlands and Islands. He also provided the
plans for the manse, according to a standard pattern (with
variations), which he drew up for the dignity and comfort of
pastors in remote places: there are forty-three of them
altogether. One cannot travel far in Scotland without coming
across the work of this remarkable engineer and architect,
who began life as a humble stonemason in Dumfriesshire, was
entirely self-taught, and acquired a huge range of skills by
prodigies of application. He worked on the building of Somer-
set House, where his genius was spotted by Chambers; but he
seems to have acquired his knowledge of engineering entirely
by practical trial-and-error. This was characteristic of the way
the Industrial Revolution took place, without the smallest
assistance (indeed, often against the determined efforts) of the
products of the great public schools, and of Oxford and
Cambridge. It was Telford, on behalf of the government, who
surveyed the possibilities of opening up the Highlands, forced
through his recommendations, and supervised their execution.
He built the Caledonian Canal and, more important, the first
road-system in Sutherland and Caithness. No man in history
did more to raise Highland living-standards. He spanned the
two cultures, for he was a poet and artist as well as a scientific
engineer. It would be hard to equal the beauty of his functional
designs, which survive to rebuke the sterile ugliness of our

glass and concrete boxes, and to remind us that efficiency and
taste can be married. I salute him as a great Scotch polymath
and public benefactor.

There is nothing that a reasonable man could complain of in
Iona. It has a street of plain and tidy houses, no noise or dirt
or cars, small, rounded hills, good pasture, friendly cattle and
sheep, innumerable magic wells, clean, deserted beaches. Yet
we had no sooner surveyed the place, and inspected its points
of interest, than we wanted to get off it, and began to fuss
about the ferries. We were suffering from a sense of *insulation*,
which I think lies at the heart of the migration from the
Hebrides. The feeling that there is no firm earth on which one
can walk away, that one is entirely dependent for movement
on the vagaries of boatmen, the exigencies of the timetable,
and the fortunes of tide and weather, induces a sense of
deprivation and the dread of being marooned. There is no
rational explanation for this emotion. Even if George and I
had been stuck on the island for a few hours, or even for a day
or more, there were plenty of things we could have done with
our time: made sketches, for instance, or written up our notes
or read the many books about Scotland we carried with us.
But we wanted to get away; we resented the loss of freedom
insulation brings. Moreover, these islands can become quickly
oppressive, because they emphasise the poverty of untamed
nature. Dr Johnson said they consisted of nothing but 'water
and stone'. This is not strictly true, but one sees what he
meant. There is a sameness of texture in the smaller islands
which numbs the senses: no corners to turn, no expectation of
any variety. When Johnson and Boswell were marooned on
Coll by bad weather, the Doctor occupied his mind by writing
geometrical formulae in his pocket-book, a good example of
his sturdy academic heroism. But even he soon became
depressed, and let out the anguished cry of the marooned: 'I
want to be on the main land, and go on with existence. This
is a waste of life.' To be on a small island is, indeed, a waste
of life, and we were glad when the ferry-boat turned up.

CHAPTER THREE

*The hotel that runs an airport – Why Tobermory is not
St Tropez – The lairds and the Galleon – A Presbyterian service – How the Kirk became soft – Dervaig
and its silence – The horrors of Fort William –
Apartheid in the Highlands – Old stones and new
money – The terrifying Pass of the Cattle – Applecross and religion – The black hills of Skye.*

One of the striking things about northern Scotland is the
abrupt and seemingly inexplicable changes of vegetation,
ecology, even climate. From the bleakest moorland you turn,
without explanation, into valleys of Mediterranean richness
and profusion. Driving around Mull, we went from barren
peat-bogs into dense forest and out again, from lunar landscapes of broken rocks into green pastures and cattle-ranches.
There is a terrifying road around Loch Na Keal where huge
boulders impend above, and rocks are liable to tumble down
any second: the surface is littered with stones. But only a mile
away is the glittering, pastoral island of Inch Kenneth, where
Dr Johnson was entertained by Sir Allan Maclean; he found
'unexpected neatness and convenience' in Sir Allan's house,
'all the kindness of hospitality, and refinement of courtesy'.
This island would suit me very well. It does not have the
remoteness of Iona. One could row to the main coast without
difficulty, and there is a good, solid, three-story house on it,
whose owner I envy. There is also a ruined chapel, of great
age, where (as we now know from his private diary) Boswell
went secretly at dead of night to pray, as he thought, to
St Colomba. But he was confused by his ignorance of Latin,
and ended up praying to Christopher Columbus.

Then we went into the pine-woods, and eventually found a
little hotel built of logs. This hotel is the only one I know
which operates an airfield. We were having lunch when a bell

rang, and the waiter and the kitchen-maids ran out to man a little red fire-engine which stands by the side of the airstrip. The airstrip, I must say, did not inspire me with confidence, for it was nothing more than a field, and appeared to slope down towards the sea. But the aircraft landed without difficulty, and a surprising number of people, and pieces of baggage, emerged from it. The control-tower was the reception-desk at the hotel, operated by a grey-haired lady, who cheerfully answered the phone-calls from grand, professional airports in Belfast, Glasgow and elsewhere, and so co-ordinated the traffic. Supposing, we asked, two aircraft wished to land at once, would she 'stack' them, as they do at Heathrow? 'It hasn't happened yet,' she said. I reflected, not for the first time, that much of the elaborate mystification which surrounds the work of great international airports is pure stage-setting, part of the bunkum which persuades the public to accept the regimentation, inconvenience and expense of air-travel. I once asked a jet-pilot to explain exactly how he navigated a landing at London airport. 'Oh,' he said, 'I just look out for Southall Gas-Works.'

When we reached Tobermory, and asked about the ferries to the mainland, our sense of insulation became acute again. We were told, with what seemed to us sinister relish: *'Ye'll no get off the island on a Sunday.'*

Tobermory is the prettiest place you can imagine, high and dramatic, tumbling down into a deep, circular bay, and guarded by woods, islands and mountains. Our hotel was perched on a wooded cliff, and the view I sketched from my window beat anything I have seen in those famous Cornish and Devon resorts. 'An excellent harbour,' said Boswell. 'Such a variety of people, engaged in different pursuits, gave me much gaiety of spirit.' Johnson was more prosaic: 'The port had a very commercial appearance.' The commerce has largely gone: the coming of the railway to Oban saw to that. But there were many yachts in the harbour, and in theory at least nothing but the weather prevents this place from becoming another St Tropez. The little town, like many in the far north, was built all of a piece; the high houses fronting the bay are

delightfully irregular within a uniform scheme, and smartly painted in black and white. It cheers you up just to look at it. But theory, in these parts, breaks down on the adamantine rock of the kirk. There is here a relentless effort to restrict enjoyment. You could not even buy petrol on a Sunday. The *Shorter Catechism* asks: 'How is the Sabbath to be sanctified?' and answers: 'The Sabbath is to be sanctified by a holy resting all that day, even from such worldly employments and recreations as are lawful on other days; and spending the *whole time* in the public and private exercises of God's worship, except so much as is to be taken up in the works of necessity and mercy.' Now Tobermory follows this commandment literally, at any rate to the extent of stopping people from breaking it. So we could get no boats to take us to the islands on a Sunday, and the organisation of any anti-Sabbatical activity seemed to present insuperable difficulties. Tobermory, we can safely predict, will not become another St Tropez.

Not that the islanders despise money. But, like Irishmen, they live in hopes of getting it all at once. The lairds, especially, are always hoping to find a pot of gold at the end of a rainbow. There is not a laird in these parts who does not expect, some day, to find himself sitting on a diamond mine, or a huge uranium deposit, or even an oil-field. For centuries they have been fertile in schemes to get rich quick. Often they lose their money in crazy ventures. Arthur Balfour, for instance, though a sceptical Prime Minister, and the author of *Philosophic Doubt,* spent a quarter of a million on a fruitless attempt to turn peat into powdered fuel. The Laird of Mugg, in Evelyn Waugh's *Officers and Gentlemen,* was typical – indeed I have always assumed Mugg was based on Mull – for he hoped to make his fortune by the skilful employment of high explosive to exploit his island's resources. Now Tobermory Bay exercises a special fascination for these lairds. For it is known, or at any rate believed, that a Spanish galleon, containing all the Armada's cash, sank there in 1588. The story is that the ship put in for repairs and stores, which were provided, and then refused to pay for them; not very probable, one might think, if it is true it had thirty million ducats aboard. Anyway, Donald Glas Maclean was sent on board to get payment, was imprisoned there, but contrived to blow up the powder magazine and sink the ship at the entrance to the bay. Now for some reason which I do not understand, but is doubtless connected with the complexities of Scotch land law, the ship belongs to the Dukes of Argyll; and ever since the beginning of the eighteenth century they have been trying to get at it. An expert survey made in 1968 confirmed that the wreck is there; but whether there is treasure in it, and if so how much, no one can say. The lost galleon of Tobermory Bay is very like the burial treasure of Alaric, which the people of Cosenza are convinced, to this day, lies under their river bed, as related by Gibbon. I have now inspected both places; and would not invest a penny in either. But I am sure the lairds will continue to ferret for their ship.

I was determined to go to a Presbyterian service. Since a misguided Pope destroyed the Roman Catholic mass, and turned an ancient ceremony, conducted with uniform elegance throughout the world, into a polyglot shambles, I have had no scruples about attending churches of any denomination. I am as happy worshipping with the Congregationalists, or the Jews, or Anglicans high, low and broadbottomed, or Christian Scientists, or even Mormons, as I am with papists. But Scotch Presbyterianism was new to me. George agreed to accompany me with a certain degree of fascinated dread. For he had been brought up a Presbyterian; had been forced to attend throughout childhood and adolescence until, at the age of 16, he had stormed out in protest against an anti-semitic sermon; and had never been back since.

The kirks of Mull offer a wide choice. One, indeed, advertises 'a good, old-fashioned Scottish service'. Our kirk, I should say, was pretty middle-class, to judge by the congregation: black suits, stiff collars, white and pale violet gloves, lavender, and floral hats. The hymns were good: rousing, Biblical words and sound Scotch tunes. I sang them lustily, though George, who was twitching a bit, stayed silent. There were other twitchers in the congregation, mainly female; perhaps it is a characteristic of Presbyterian worship. The children had an instruction period of their own, and a sermon, and were cross-questioned by the minister, not very severely. Then they were marched out – not, as I supposed, to freedom but (George said) to Sunday school, poor things. The rest of us got one hour and twenty minutes, and I must say I was disappointed by the sermon. It was well-phrased, evidently prepared with care, and distinguished by spectacular gestures, of a kind which would certainly seem odd in a modern English church. But there was no Hell-fire; nothing which would shock, or excite, or anger members of any other Christian denomination, and a lack of theological subtlety. Nor was anyone called to the stool of penitence. Indeed I saw no stool, or anything resembling one.

Perhaps the kirk is becoming soft. It used to be the fiercest in Europe. Like the sixteenth-century English, the Scots once believed that they were the chosen race, and their church a special divine instrument. There was 'a verrie near paralel betwixt Izrael and this Churche, the only two sworne nations of the Lorde.' Curiously enough, the Scots did not kill people on religious grounds in the sixteenth century; this was something they learned from the English, and practised under the Commonwealth. But they burned witches with unrivalled ferocity and took a harsh line with backsliders. On Sundays, official 'searchers' patrolled the streets, and took the names of those not at kirk. At Glasgow in 1652, the kirk-session appointed paid spies to report on lapses of all sorts by members of the congregation. I was not surprised to learn that at Tobermory absentees were fined. In 1650, a man in Culross confessed to 'sitting in his owne hous the whole tyme' of worship, so presumably the searchers had the right of entry too. People were fined for sleeping in church, and women forbidden to wear plaid headdresses, as these made hidden slumbers possible. On the other hand, people could, and did, correct the clergy, and exhort worshippers to 'pull the minister out of the pulpot be the luggis'. Sexual offences aroused the greatest severity, presumably because the fact of their commission could be proved more easily than in the case of more serious sins, like pride or avarice. An adulteress was punished on the stool of repentance every Sunday for six months. According to Professor Smout's *History of the Scottish People* 1560–1830, kirk discipline led to an increase in infanticide and homosexuality. But homosexuals, men and boys, if detected, were burned alive, 'sometimes two together'.

The kirk condemned schoolmasters for insufficient use of the tawse. It even sought to make blaspheming a capital offence. But its real power depended on the acceptance of theocracy, that is the willingness of the State to endorse ecclesiastical punishments. So long as an excommunicate was denied by law the right to hold civil office, or even to possess property, the kirk wielded enormous authority. But theocracy was a casualty of the Union. The English have always repudiated the right of clergymen to exercise secular power. In 1712 the Toleration Act forbade Scotch magistrates to enforce spiritual penalties, and the goose of the kirk was cooked. This is a point I always stress when arguing with agnostic Scot-Nats.

We spent some time in the north of Mull. There are impressive rock-formations, known as the Gribun, or Griben, Rocks, a famous resort of geologists, for they include among other

curiosities an entire fossilised tree. At Calgary we found an
immense and placid bay of silver sand, enclosed by rocky
headlands and with a prospect of islands shimmering on a
deep blue sea. It looked as though it had been designed for
Cecil B. de Mille. But then the reader must remember that the
sun shone from four in the morning until eleven at night; at
least when we were there. The locals said it was always thus,
or nearly always. The idea that Mull was a rainy place was a
malicious fiction, put out by enemies to scare away tourists.
If this is so, why do so many people wish to come to Mull
nonetheless, and why do the people of Mull make so little
effort to accommodate them? The car-ferries, especially, are
inadequate to handle the summer traffic; and hotels seem to
take delight in telling would-be guests that they are 'full up
till the third week of September' – it is always the third week.
There is something listless and unenterprising about the Mull
folk. We went to Dervaig, a tiny village with a pencil-steeple
to its kirk, and pretty, one-story cottages. It seemed to
acquiesce in its decay, and the departure of youth. The locals
blame the lairds for the failure to exploit Mull's attractions;
and it is true that the lairds do not, on the whole, welcome
visitors, who are liable to disturb their shooting. But the
lairds sometimes have justice on their side. A local en-
thusiast, discoursing on their iniquities, told us that *they even
opposed the plan to hold a car-rally in Mull.*

We were up at 5.30 to catch the early ferry from Craignure
to Lochaline on the mainland. Few people go this way, but it
avoids many miles of busy roads and takes you through the
lonely glens of Morven. Early on a summer's morning, this
huge territory seemed uninhabited; not a house or a face did
we see. But there were animals everywhere, and mostly they
sat in the middle of the road. First we disturbed a sheep and
two lambs; then two wild-looking cats; next, a family of
rabbits; then hedgehogs; then cattle; then more sheep, and so
it went on. All seemed reluctant to move. All returned the
second we passed. Why do these animals like roads so much?
Is the grass on the verge sweeter? Do they enjoy the smell of
petrol, or need human company?

Our route skirted Loch Sunart, on which stands a little village called Strontian. The Scots admire this village because, when the established kirk split in 1843, its inhabitants seceded unanimously, defied the local laird, and when he refused them land to build a kirk, made one out of a ship and anchored it offshore, where they held services in the teeth of his rage for thirty years. This is the kind of thing which ought to happen in the Highlands more often. On the other hand, Strontian has a sinister ring, too; for in the eighteenth century the local mines produced an element which Sir Humphry Davy isolated, and called strontium. And we all know about strontium-90. Glen Tarbert, above Strontian, is likewise sinister: a fine, dark, silent and menacing glen, which opens up into a majestic panorama of the highest mountains in Scotland, seen across the foreground of Loch Linnhe. I recommend this route, which takes you to a tiny ferry-crossing at Ardgour, rather than the more popular one on the eastern side of the great divide. You are quite alone. There are no cars or people. And you can see the great peaks of Glencoe and Nevis in their true perspective, across sheets of calm, deep water.

Unfortunately it is less easy – indeed it is impossible – to avoid Fort William, that blot on the Highlands. Towns are so rare in this part of the world, and so welcome in many respects to travellers from the wilds, that one is prepared to make allowances. But Fort William has been wholly given over to the philistines. General Monk's men first set it up as a base in 1655, and in the eighteenth century it became the military key to the western Highlands, for it bestrides all the communications. It ought to exhibit the best virtues of Georgian military architecture, which reached their finest expression in the Highlands: a smart and classical fort, a fine governor's mansion, neatly-planned streets, and well-built houses for the officers, revenue men and traders – all those relics of the colonial-frontier days which so often astonish and delight the eye in the far north. But it has none of these things. The fort was pulled down to make way for the railway in Victorian times; and such Georgian houses as remain are buried beneath a welter of signs and commercial impedimenta. The narrow street is choked with traffic; and the new buildings which have been put up in unrestricted profusion – hotels and motels, shops, garages and boarding-houses – are of unspeakable ugliness. It is a town to pass through as quickly as possible: no easy matter, for it forms a bottleneck on all the roads of a vast area.

Shaking the dust of Fort William off our wheels, we drove towards the sea again, and stopped for a drink at a lonely hotel. Now here was a curious thing. There was a cocktail bar in the hotel, which George disdained; and there was a public bar, quite separate, down the road. This bar was full of hydro-electric workers, and other brawny men; there was a good deal of noise, and drinking of pints of beer with whisky 'chasers'. But it was a cheerless place, a most miserably furnished room, chill and badly-painted, a bare den for the consumption of liquor. As George pointed out, it was like the beer-shops in the native townships of Southern Rhodesia and South Africa. The lower orders are permitted to drink, by gracious favour of authority, but no more. They must have no comfort, no sense of welcome; just drink up and move on. Many hotels in the Highlands operate this system of social apartheid. The rich are provided with comfortable chairs, and lights, carpets, fires and attention. The workers, who pay almost as much for their drinks, get little more than a shed, which might have been expressly designed to induce a sense of sin. Here, as so often in the Highlands, the kirk and the lairds work together, though with different objects. The kirk cannot actually prevent the consumption of liquor, much as it would like to. But it contrives an atmosphere in which men take drink in a squalid and furtive way, as a form of sottish indulgence. Moreover, no woman would willingly enter such a bar as this so there is a separation of the sexes, too, and the men drink more than they should. This suits the kirk's book, for they are anxious to divide people into the elect, who do not frequent such places, and the damned, who make beasts of themselves there. As for the lairds, though they can no longer force the Highlanders to touch their bonnets, or even prevent them from earning high wages from public authorities, they at least try to preserve rigid social distinctions on licensed premises.

The nasty Scots public bar makes the Highlander seem what the laird wants him to seem: a different sort of animal. It is a form of racism, wholly unsupported by law; and I do not know why the working-men do not invade the lairds' bars, and demand their rights. But the Highlanders, who exhibit prodigies of physical courage, are timid when it comes to moral gestures. That is why the lairds still crow triumphantly on their dunghills.

Some lairds, of course, have survived by marrying commercial money. On wild Loch Duich we saw one of the showpieces of the Highlands: Eilean Donan Castle, built on an islet linked to the mainland by a fine stone bridge. The castle was

the headquarters of the MacRaes. It must have been a formidable strongpoint and, garrisoned by Spaniards, Irishmen and other recalcitrants, it gave the English a good deal of trouble long after the 1715 rising was put down. In 1719, three Royal Navy battleships (or what the guide called 'the full might of the English navy'), sailed up the sea-loch and blew it to bits. The laird-like man who showed me round the castle seemed to consider this an outrage. But what were the English supposed to do? The Stuarts themselves were no friends of the Highlanders; it was they who invented the Commissions of Fire and Sword, the normal method of royal 'peacekeeping' from the fifteenth to the seventeenth centuries in this part of the world. Moreover, these northern clans were often organised thieves and murderers, notorious for their treachery and lack of mercy. Only by cowing their taste for violence was it possible to civilise the Highlands and raise the living standards of the people, as the Campbells of Argyll were the first to recognise. The Campbells still have a bad name in the Highlands; I even met a man who said he 'wouldna sell land to a Campbell' (I doubt if he had any land to sell). But it seems to me that they were usually on the side of common sense and the rule of law. The Macdonalds of Glencoe, their victims, have been hugely romanticised; but they were the most thieving and murderous tribe of the lot. I do not say it was right to massacre them. But I doubt if their disappearance caused much unease to their neighbours at the time. Similarly, many sensible and peaceful Highlanders must have been delighted to see the Navy sort out the MacRaes.

Anyway, the castle was demolished, and remained so for two centuries. Then a MacRae married what my guide termed 'a maltster heiress', one Miss Gilstrap. He changed his name to MacRae-Gilstrap, and Gilstrap cash was poured into the rebuilding of the castle. In the 1920's over a quarter of a million pounds, in good, old-fashioned English money, was spent on reconstructing it just as it had once been, and there the MacRae-Gilstraps lived in style, amid tartans, sporrans, bagpipes and so forth. Whether the money was well spent, it is not my business to say. But certainly the castle gives a lot of innocent pleasure to modern children, and is an excellent subject to sketch. Miss Gilstrap's portrait hangs in the dining-room, and she struck me as a formidable lady, looking in Evelyn Waugh's phrase, 'every inch of her income'. 'Oh, what a nice, kind face she had,' said a woman-tourist; and the laird-man soberly agreed.

The MacRaes came from nearby Glen Shiel. The Gilstrap marriage was not the first time they had contracted an alliance of utility, at any rate according to Dr Johnson. They were originally 'an indigent and subordinate clan', the servants of the Maclellans. But the Maclellans were wiped out under Montrose, and their women, wrote Johnson, 'deprived of their husbands, like the Scythian ladies of old, married their servants, and the MacRaes became a considerable race'. Shrewd and fortunate MacRaes! Their glen is a fine one, fringed by formidable peaks on either side. Here, resting on a rock,

Johnson first conceived the idea of writing his *Journey to the Western Islands*. This rock, or anyway a rock, is shown to visitors, and called, not surprisingly, 'Johnson's stone'. So I sat on it. The mountains rise to over 3,000 feet, and are called by misleading names, such as the Five Sisters of Kintail, the Saddle, and so on. Dr Johnson rebuked Boswell's easy acceptance of these silly names; and when Boswell called one hill immense, said: 'No; it is no more than a considerable protuberance.'

I first heard of the village of Applecross in a book called *The Charm of Scotland* by John Herries McCulloch. Despite its off-putting title, this little book is full of information about some of the remotest places in Scotland, which the author explored during many years working as a reporter on the *Scottish Daily Express*. But even he found Applecross difficult to get to, or rather out of, and I think it can fairly be given the prize as the least accessible spot in the British Isles. To begin with, it is the only real community, and on the farthest side, of a vast area of mountain and high moorland where nobody lives at all, to the west of Loch Kishorn. A little boat puts in there once a day in the summer, up from Lochalsh – that is, when the weather is good enough. Otherwise, the only way to get there is by a single-track road which winds over the highest pass in Britain, and is usually blocked by snow in winter. It is the sole entrance, and the sole exit. At the foot of it, on the mainland side, a formidable official notice utters dire warnings, and concludes: 'Not advisable for learner-drivers' – something of an understatement, I should say. I would not recommend the 'Pass of the Cattle', as it is called, to those who are awed by the terror and majesty of unbridled nature.

The trouble with this pass, as we discovered, is not just its height, and the narrowness of the road, and the vertiginous hairpin bends, but the tremendous sense of oppression which the huge black cliffs of the mountains produce on the traveller, who feels tiny and unprotected in his fragile metal box. We had already seen, a few days before, what can happen to these metal boxes if they leave the road in a Highland glen. Near

Inveraray we had seen far below us, on the grass, a car which was being anxiously inspected by its occupants. It had fallen over the rim of the road – no doubt while its driver was taking a quick look at the scenery – and bounced to the bottom in a series of steps. The angle was not steep, but the effects were dramatic. The car was exactly like a broken toy which had been stamped on by an angry child, with the chassis protruding through the bright new paint of the bodywork: beyond any doubt a write-off, and its passengers lucky to be alive and unhurt. Another traveller, who had gone to have a look, came panting back up the slope in a state of great hurry and excitement.

'Are you going to get help?' we asked.

'Good God no,' said the man. 'I'm going to my car to get my cine-camera. I've got coloured film in it, too. This is going to make my holiday!'

Thus the climax of a remarkable home-movie was born.

Anyway, the Pass of the Cattle makes Glen Sheil, or any other Highland glen I have seen, look pretty tame stuff. There, a slip off the road would certainly be fatal, and the driver is painfully aware of it. The sun blazed, the sky was blue, but we appeared to be ascending into darkness and gloom as the cliffs closed in. At the top, indeed, a black cloud appeared abruptly from nowhere, and rain splashed on to my sketchbook as I made a quick drawing. But the vast and complex panorama was really beyond my skill. One looked down into a series of immense *cirques*, as the road vanished from one rocky platform in a swoop to the next; and beyond it all, miles away, beams of bright sunlight flashed on Loch Kishorn and the mountains of the interior. It was a scene for Turner at the height of his powers, or for James Ward, whose terrifying masterpiece, 'Goredale Scar', hangs in the Tate Gallery. Now Ward made Goredale look more formidable than it actually is. Here the reverse was true, for I could not convey the fear and desolation which the scene inspires.

On the reverse side of the great mountain barrier, we found the road much easier, crossing high moorland and then falling in graceful loops to the lonely coast. Once again, there was a sudden and dramatic change in climate, scenery and atmosphere. Within a few minutes we passed from a Scandinavian mountain desert down into a Mediterranean littoral. Fresh green woods raced towards us, and there was the scent of a multitude of wildflowers. The waves danced in the sunlight, and we were in the safety of Applecross, which leads its remote existence perched on the rim of the wildest mountains in Britain.

All inhabited places have a business. The business of Washington is government and crime, of Coventry cars, of Essen steel, of Salzburg opera, of Tangiers vice, of Kuwait oil, of Calcutta poverty, of Zurich money. In Applecross, the business is religion – though no one, I hasten to add, makes a penny out of it. There is a little hotel there, where the beds are soft, and the food wholesome, and you can send out for a bottle of wine to the licensed grocer next door (no nonsense about twelve hours' notice, either). But otherwise there are just little houses and churches. Obviously, the people have to make a living, and do so – I imagine with some difficulty – looking after sheep; certainly, sheep and lambs lie in the one road all day and night, with no one to disturb them. But any energy left over from making a living is devoted exclusively to the business of religion, in its purest Presbyterian form.

Why should Applecross be a centre of religious fervour? So far as I have been able to discover, it was not visited by St Colomba. It was not the ancient site of a monastery or hermitage. Some people, it is true, say it got its name because a primeval landlord planted five apple trees in the shape of a cross – a likely tale, indeed! Others, that apples grown there bear the sign of the cross; but they manifestly do not. Scholars of Gaelic place-names say it does not mean apple, or cross, but comes from *aporcrosan*, that is, the most northerly rivermouth in Scotland. But it is not the most northerly rivermouth in Scotland, and never has been, as any man with eyes can see.

In any case, in not very remote historic times, the people of Applecross were regarded as barely Christian. After the Reformation, many places in the Highlands and Islands fell through the net of organised religion. When the Roman

Catholic administration was destroyed, no regular Protestant system replaced it. In some places, notably Barra, South Uist, Glenmoriston, Avertarff and Glen Garry, Jesuit missionaries were active, and the people there reverted to Roman Catholicism, which they still profess. But in other remote parts, contact was lost, and strange things happened. In 1656, the presbytery at Dingwall, near Inverness, reported that the people of Applecross, and nearby, had formed a semi-pagan cult of their own, worshipping St Mourie, or Mastrubha, some ancient, and perhaps mythical, holy man who gave his name to Loch Maree. They were said to be sacrificing bulls and adoring stones.

Obviously, the people of Applecross liked to be different, and rather special. Indeed, anyone who takes religion really seriously is likely to find it hard to fit into any universal cult. So many interpretations of the Bible are possible; so many nuances of dogma and morality will strike a strong-minded and thoughtful person striving for the exact truth, that it is virtually impossible for him to find anyone else in perfect agreement.

Happily, Presbyterianism is well suited to this approach to religion. By its nature and organisation it tends to disintegrate into active and self-contained fragments. It is made for remote and scattered places, peopled by determined individualists. Even before Presbyterianism reached the Highlands, in the late eighteenth century, it was found impossible to maintain an established Scotch church intact, partly because, under English influence, the system of lay appointments to benefices had been made illegal. In addition to the fissiparous sects which had branched off in the sixteenth and seventeenth centuries, by 1740 there were four more: the Old Licht Burghers and the New Licht Burghers, the Old Licht Anti-Burghers and the New Licht Anti-Burghers. And there were more to come, until in 1843, the established church itself again split, with the Wee Frees repudiating lay advowsons and striking off on their own. Even when, in the 1920s, some Wee Frees and the official kirk made it up, others stuck grimly to their independent position outside the established fold. So there are three churches of Scotland, quite apart from the

Episcopalians, or 'Piskies', and I don't know how many offshoots from all of them.

This, I imagine, is why Presbyterianism took such a strong hold of the Highlands when the clan system broke down after Culloden. The clansmen had gone to war on their chiefs' orders, often unwillingly; had been defeated, and were then deserted. Not only this, but the chiefs, finally accepting the southern rule of law, began to take up sheep-farming, and throw the clansmen off lands they believed to be theirs by ancestral right. In this mood of bitter disillusionment, betrayed by their natural leaders whom they had regarded as fathers, the clansmen looked for a new social and ethical system, and found it in the kirk, or rather in a series of kirks. These provided a hierarchy to replace the one that had gone, in the shape of ministers and elders; the exclusiveness of small, localised and inter-related communities; a focus for fierce loyalties, an arena for argument – in short, something to fill a yawning vacuum in their lives. The clansmen had, above all, respected purity of lineage; this had now been dishonoured, indeed made meaningless. So purity of doctrine became an exciting and formidable substitute.

We came to Applecross at communion time, which takes place only once a year, and over the space of a week. The men wear black suits for it, and it is a matter of fine judgment, and anxious discussion, whether they, and the women, are worthy to take communion at all. Their lives during the past year are mulled over, and the verdict (often unfavourable) pronounced. No wonder they looked so worried, though the people of Applecross are exceptionally law-abiding and worthy, for excommunication, or even deferment of communion, is a fearful punishment. But of course there is more than one church; and it is possible to shift from one to another. Which is the most severe, and the most exclusive, I do not know. Even George's experience as a Presbyterian proved unequal to this problem; and to an untrained eye like mine, the differences seem almost imperceptible.

There is the Church of Scotland, presumably the established one, which advertises 'Gaelic 12 noon, English 6 pm'. Then there is the Free Church of Scotland ('English and Gaelic as

48

required'), presumably the Wee Frees. But there is also the Free Presbyterian Church of Scotland, a mystifying name, for all three are unmistakably presbyterian. Anyone anxious to detect crucial nuances in the service will find it difficult, for they are held at exactly the same time, on Sundays and Wednesdays. All three churches look alike, inside and outside. Moreover, there is another church, bigger than the rest, which has been beautifully restored, evidently at great expense. But this, though plainly presbyterian too, gives no indication of its sectarian standing. Perhaps it has an ecumenical purpose: it is certainly large enough to contain all the local inhabitants. There is no sign of open rivalry between the sects, and no evident proselytising. All cater for the elect; but some are more elect than others. A believer in Applecross presumably finds his own level in the course of time. But the process of rejection can be shattering. Some find themselves gradually pushed from one church to another, ending up as a papist, or a pagan, or an atheist – all three, to a true Presbyterian, much the same thing. Such intensity of religious life causes great unhappiness to some; but to others it brings both excitement and security, a dramatic sense of purpose in life and the prospect of eternal bliss – things hard to come by in modern British society.

We found a beautiful and rich girl in Applecross, who had driven over the terrible pass, trembling, in a fast and powerful sports-car. She was known as Mrs ——, but had discarded her wedding-ring. Her face was sweet and sad, and her hands rested in her lap in a gesture of resignation. She had fled from some personal tragedy to seek peace in Applecross; and we hoped that she found it.

Just beyond the last houses of Applecross, we followed the road which curves upwards, into a region of low hills and lochans. There we sat down to sketch the view out to sea. Applecross is indifferent to the high land behind it; its links are seaward. In winter the mountains cannot be crossed. But occasionally the little boat fails, too. A girl who helped at the hotel had been due to get married. The ferry, however, proved too small for the wild seas. It could not get her off. So

MacBrayne's put on a big ship, which took her to her wedding, proving that the hearts of Messrs MacBrayne's are not made of stone. From this place we looked across the low island of Raasay, where Johnson said, 'We found nothing but civility, elegance, and plenty,' and where he and Boswell were entertained to a dance every night. Boswell was informed that in Raasay it rained nine months out of the year. This we could well believe, for the island lies under the lee of Skye, whose mountains loom over its low profile. In my experience, there is nothing to beat Skye for rain. I once attended a climbing course there, in the Black Cuillins. For a fortnight, I toiled up and down these rocky mountains, which are the fiercest in the British Isles, made of rough gabbro rock, which tears the skin off your hands. Not once did I see the tops from the bottom, or the bottom from the tops, and it did not merely rain, it poured the whole time. From the hills beyond Applecross, we watched the Cuillins make the bad weather. Above us the sky was perfectly blue; and beyond Skye itself all was sunlit. But every few minutes a black cloud descended from the heavens, settled on the summits, emptied its deluge, and then disappeared into thin air. The shapes and colours of these peaks changed from second to second, so that the sketch I drew represents not a moment of time but the distillation of innumerable movements, which raced ahead of the pencil. Then we left Applecross, surmounted the terrible pass in safety, and turned to the far north.

CHAPTER FOUR

*In a Highland nature reserve – Recreating the primeval
forest – Does the Gulf Stream exist? – Ullapool and
the rain – The lady from the Yorkshire moors –
The finest mountain road in Britain – Fish and the
red-faced man – We are thunderstruck – Scottish
female indecency – The Cape Wrath ferry – Arts and
crafts in Durness.*

We now entered the finest part of the Highlands, where few
people live, the mountains are at their wildest, and the
conservationists, for once, are a step ahead of the developers.
It is true that the roads are being 'improved': a two-lane one
is replacing the old ragged track along Loch Maree, and it is
equipped with viewpoints, picnic-places and other doubtful
amenities. It is also true that southerners, or lowlanders, or
at any rate 'foreigners', are buying themselves holiday-homes
in these parts. At Shieldaig, we were told, a croft-house or a
cottage, if advertised only locally, would rarely fetch more
than £700; but if advertised in the London papers would bring
up to £3,000. This posed a dilemma for the people. They did
not want, they said, well-to-do strangers moving in; on the
other hand, why should a local man sacrifice £2,000 or more,
in realising the value of perhaps the only property he possessed? Obviously, the battle for solitude will be lost. But it
has not been lost yet, and powerful public bodies, such as the
Nature Conservancy, are fighting a vigorous rearguard action.

We drove up Loch Torridon and through the great glen
which connects it to Loch Maree. Here, in 1951, the Ben Eighe
nature reserve, the first such in Britain, was opened. It covers
over 10,000 acres, where golden eagles fly, and pine-martens,
wildcats and 150 red deer, as well as many roe deer, live in
safety. I say 'in safety'; but the sad fact is that deer, un-
molested, breed rapidly, and one in five have to be 'culled'

every year to preserve the balance, and protect the weaker ones from winter starvation. Moreover, the object of the reserve is not just to protect animals but, above all, to get the ancient Caledonian forest to breed itself. Some fragments of this forest, which once covered vast areas, and perpetuated itself, still remain. What the experts do not know, and what they are trying desperately to discover, is why the forest, now protected, does not automatically spread itself. Of course it would help if they knew what had happened in the past. Is it true, as local tradition maintains, that the Vikings burnt it down? Or was there some other momentous change? If the trick could be found, these vast areas might suddenly become sources of great wealth. A forest should be a dynamic thing, and usually is. In the Lake District, the native trees spread naturally, if only people would let them. In E. M. Ward's *Days in Lakeland*, the best book ever written about the district (now, of course, out of print), he justly remarks: 'If modern man found himself suddenly powerless to interfere with the trees, the forests of High Furness and Coniston would, slowly but quite certainly, close in upon Hawkshead and Coniston towns, these alien growths, and smother them.' Why does not the same thing happen in the Highlands, when man holds back? This is what they are trying to discover at Ben Eighe.

Other simple but dramatic solutions continue to elude man in the Highlands. If only a universal and profitable use could be found for the heather, or the bracken, or the peat! Then the Highlands would be rich, overnight. As it is, virtually every form of enterprise there is subsidised, in one way or another, by southern taxpayers. Distance, altitude and rain make business, or even mere existence, impossible without public money. The whole land is, in a sense, one vast nationalised industry, for even the lairds – indeed above all the lairds – have waxed fat on the compensation they have received from public boards and the subsidies for developing their forests. In this land two great empires flourish: the Forestry Commission and the North of Scotland Hydro-Electric Board. The Forestry Commission builds its own roads and houses; is a state within a state. The Hydro-Board is still more ubiquitous, for it is empowered by statute to promote the economic and social development of the entire North of Scotland. Its achievement is immense. In 1948, when it began operations in earnest, only five per cent of the farms and crofts had electricity; now virtually all are covered, and in the meantime, by building hundreds of miles of road in remote country, vast areas have been opened up to visitors like George and myself. The Hydro-Board has brought a measure of prosperity to the north; it has done nothing but good; it has performed miracles in preserving the countryside, and its fish and game, while making it possible for locals to get work and tourists to explore. But it is hated by most of the lairds, for it operates to the benefit of all, and forms a rival centre to their power.

Moving farther north into Wester Ross, we found the contrast between the wild interior and the Mediterranean coast still more striking. At Poolewe, where the waters of Loch Maree flow into the sea, we came across a great sub-tropical garden, over 2,000 acres in extent, which has been created in the last hundred years. The guide-books said this was due to the Gulf Stream. George said that he did not believe in the Gulf Stream theory; that it was one of those geographer's myths, taught to generations of gullible schoolchildren, and accepted simply because nobody bothered to challenge its premises. How could the presence of warm salt-water nearby affect what grew on the land? Anyway, the water was freezing cold, despite the hot sun, as any fool could discover by dipping his toe in it (this was undoubtedly true). Maybe the Gulf Stream had disappeared – had anyone checked recently? Or maybe it had never existed. Be this as it may, the fact is that the headland where the Inverewe Gardens now flourish was, until 1862, a barren and rocky place, Gulf Stream or no. Then along came Mr Osgood Mackenzie, who, despite the fact that he was a laird and sportsman, was also a gardener. He put in shelter-belts of trees; he dragged up enormous quantities of rich soil in baskets; and he planted endless varieties of shrubs, trees and flowers. The garden is now, some think, the finest in Britain. It has the largest magnolia tree in existence, a fifty-foot high hydrangea, exotics from all over the world, and miles of sumptuous rhododendrons, which were then in bloom.

Mackenzie's daughter gave it to the National Trust, which keeps it open all the year round, and an average of 100,000 people come to see it. It shows what can be done in northern Scotland, by those with intelligence, imagination and energy.

On the other hand, once we turned the corner in the road by Little Loch Broom, and its bigger sister, we were back in the sinister harshness of the wasted north. Here were deep, black waters, dank and unexplored moorland, a sense of virgin territory. Does anyone ever visit Loch na Daimph, I wonder? Or Loch a Choire Mhoir? Or Strath Mulzie? Or Glen Beag? In these parts there are hundreds of thousands of acres, wholly inaccessible by road, or even track, given over to the deer and little else. You could wander off into the 'forests', get lost, die of starvation and exposure, and never be found. No one knows what to do with the interior, though sporadic attempts have been made to develop the coast.

From the heights we drove down into Ullapool, which stands on a little promontory on Loch Broom. It was founded as a colony, in 1788, by the British Fisheries Society; and the ubiquitous Telford had a hand in the planning of its streets and houses, which must once have been handsome. But the fisheries did not flourish; Ullapool failed to become what we would to-day term a 'growth point'; and it is now a rather scraggy holiday resort, which threatens to become a smaller-scale Fort William. There, almost for the first time on our jaunt, it started to rain. A pleasure steamer drew up at the pier and disgorged a huddled mass of trippers, wearing Pakamacs, and sprouting umbrellas. They had been to the Summer Isles, and had got very wet and cold. The black clouds dropped lower and lower; the loch became the colour of gunmetal; an icy wind appeared straight from the North Pole, and whipped sweet papers viciously over the glistening streets. There was the smell of fish-and-chips, and the wail of fractious holiday-children. At a general shop, I tried to buy a paper. Had they a national paper? No. Or a Scottish paper? No. Or a local paper? No. They had to be 'ordered' (like wine at the Misses Nicolson's). But didn't many holidaymakers ask for newspapers? 'Yes,' said the man in exasperation. 'They dhu. An' I haf to explain to them, time and time again, *there's*

no demand here.' Had they any Colgate's toothpaste, then?
No: only Maclean's – 'and a good Highland name, too.'
Baffled, we left Ullapool.

Next morning, at our hotel in the wilds, we shared a breakfast-
table with a couple from the Yorkshire side of the Pennines.
They were very nice, and friendly, and voluble. The lady had
one of those rich and nutty northern accents which are an art-
form in themselves, and which seem to raise the most ordinary
conversation to the level of dialogue by Mr J. B. Priestley.
She told us about herself, and when she paused for breath,
which was not often, her husband took up the tale. She had
been coming here for twenty-five years. Always the same
place. Never been abroad. It was nice here: reminded her of
the moors at home. It was a good place for walking, like the
moors. Apart from the rain, of course. And the stalking season.
And the lairds. She had it in for the lairds, she must admit.
Selfish people, weren't they? Well, weren't they? Always
stopping you from doing things. Greedy, too. If you stopped to
look at one of their trout-lochs they'd send a gillie along to tell
you not to fish. The idea. She'd never fished in her life, had she.
Well, had she? No. Not that she was against a nice trout for
breakfast. Tasty. The food here was good, she must admit.
Even in the bed-and-breakfast places. Now, here was a
shocking thing. The lairds had even tried to stop people
putting up signs saying 'Bed-and-Breakfast'. Would we
believe it? Of course not. But she'd believe it. She'd believe
anything of a laird, would she.

At this point the lady paused, and became aware of the fact
that George and I had said nothing for the past ten minutes.
The reason why we had said nothing was partly that we had
nothing to say, and partly that we had had no opportunity of
saying it. But the lady looked at us, and began to show signs
of alarm. I should explain that the expressions George's, and
my, faces assume at breakfast, indeed for a good part of the
morning, are not entirely friendly. There is nothing personal
in this; it does not reflect what we hear, or what we are
thinking. It reflects nothing at all. It is just that, at an early
hour, George's features naturally and habitually compose

themselves into the facial mask of an eighteenth-century hanging judge, who is about to sentence a seven-year-old child to the scaffold for the theft of a pocket-handkerchief; while mine, no less innocently, have a good deal in common with the Duke of Cumberland's, on the morrow of Culloden, when confronted with a chain-gang of Highland prisoners. We can't help it; we were born that way.

The lady from the Yorkshire moors took one long, agonised look at our mugs, and froze into silence. At last she said:

'I suppose I must be addressing two members of the aristocracy.'

George laughed, and I laughed, and we tried to explain matters. But the lady's confidence was not wholly restored. She eventually abandoned the aristocrat theory, but I believe that, to this day, she thinks we were two special investigators from the Inland Revenue, sent from London to put down Highland fiddles.

North from Ullapool, we drove along the finest mountain road in Britain. What makes these mountains so spectacular? It is not altitude as such. Ben More Assynt, the highest of them, is only 3,273 feet. Some are much smaller, scarcely over 2,000. But all are strikingly individual. They do not submerge their identities in a general massif, like the Cairngorms, or the hills round Fort William. They do not form a range. Some tremendous geological convulsion has turned each into a solitary sentry, glaring at its neighbours across wide valleys. They are actors in a prehistoric drama, with sharply separate roles and characters. No one who has once seen Coigach, Stac Polly, Cul Mor, Suilven and Canisp could possibly confuse one with the other. These peaks are among the oldest of the world, shattered survivors from a cataclysm 2,000 million years ago; and it is their very antiquity which has shaped their architecture. For they are made of old and hard sandstone, anchored on to plinths of gneiss, the most durable of all rocks. Thus they have overcome a tremendous battering from nature, which has swept all else away, leaving them in majestic isolation. Like huge ships, they lean into the north wind. As the twisty road winds among them, they perform a stately ballet. Sometimes Stac Polly peers between Coigach and Cul Bearg. Sometimes Suilvan raises a ragged face over the shoulder of Cul Mor. Sometimes Canisp appears without warning from round the back of Suilven. As one moves, they move too, like petrified prehistoric monsters. And they seem to watch. They are not, I should say, friendly beings.

All the same, those who travel in the far north acquire extraordinary attachments to these peaks. Powerful emotions are lavished on them. They give rise to fierce arguments, and generate quarrels. I heard one woman describe Stac Polly as 'cuddlesome'; though it, or she, looks to me like a treacherous brute. Everyone seems to have a favourite mountain, and abuses competitors. If a man likes Quinag, he will not have a good word to say for Glasven. Arcuil, or Arkle, gave its name to a racehorse, and so did Foinaven. But one man told me that, though he'd won good money on Arkle, he'd always refused to back Foinaven. Sheer prejudice, you see. Ben Loyal, as you might expect, inspired particular loyalties. George became a Ben Loyal convert, and discoursed, boringly, on its qualities, which I admit are notable. I'm a Suilven man, myself. Suilven is classic, unitary, arrogant, sublime, a natural artefact created by a transcendental and conscious intelligence: it underwrites St Thomas Aquinas's Fourth Proof of the existence of God. Ben Loyal is Gothic, sprawling, sinister, superstitious, a piece of weird imagination which has somehow gone wrong, the accidental creation of a fuddled pagan deity.

But I am writing nonsense.

We put up at an inn where they took fishing seriously. Bloodstained scales stood in the hall, surrounded by the stuffed bodies of former giants. The day's catch was noted, carted off to the kitchen, and served at breakfast the next morning. A fresh pink trout makes a good start to the day. In the evening, before bedtime, a man went from guest to guest with a list, and allocated lochs, or bits of lochs, for the next day's fishing. Men and boys sat around making flies with assortments of cotton and wire, while their wives and mothers knitted. There was much talk about the lack of rain, the heat, the sun, the north wind, the absence of midges and mosquitoes, and other deplorably inhibiting factors. 'What we need,' said a wizened, walnut-faced man, in a kilt, 'what we need, with all due respect, is a long, wet summer.'

At the bar stood a big, red-faced man, whose hair had grown white in the service of his ego. Occasionally he talked at people, in a loud, authoritative voice. Less often, he paused while they made their replies, and then continued his monologue. George put him down as a distinguished doctor; and certainly he had the combination of ignorance and self-confidence one associates with that profession. He could well have spent his lifetime bullying nurses, interns, matrons, patients and so forth. Anyway, he had now exhausted the topic of fishing, and he addressed the company at large.

'Don't talk to me about tied cottages,' he said.

No one had talked to him about tied cottages.

'Don't talk to me about tied cottages,' said the red-faced man. 'Friend of mine, senior superintendent of a group of hospitals, just retired. Had to give up his house, of course. Did he go whining to Harold Wilson? No, Sir, he did not. It's the same in the army. Officers' quarters. When your number's up, out you go, no argument. Same in any profession. But your estate workers want to have their cake and eat it. Act of Parliament, indeed! Take the average prime minister. Not a breed I care for myself, but when Wilson got his marching orders he didn't expect to hang on to Downing Street, did he? I'll say that for him. Now, take the average French prime minister.'

'The French are different,' said somebody.

'I don't deny it. Of course they're different. I'll tell you something else. The French have a new prime minister every ten minutes, don't they? That is, they certainly used to. So one says to another: "Look here, Jack, or Jacques, when we get thrown out, let's vote each other pensions, and then there'll be no difficulty." And they did. Substantial pensions, after ten minutes. De Gaulle didn't put a stop to it. Pompy-what's-his-name didn't put a stop to it. All in the racket, you see. Don't talk to me about tied cottages.'

'I'm a short-back-and-sides man, myself.
All I say is, Thank God for wee Enoch!'

No one said anything to this. Then a fat man, in a tartan-check coat, said: 'I'm a short-back-and-sides man, myself. All I say is, Thank God for wee Enoch!'

Next day we bowled along the high, straight road which leads to the farthest north. The sky was the deepest blue, though battalions of clouds marched swiftly across it. The wind was noisy, and icy, screaming down from the Arctic. The whole countryside, white and grey rock, rolling green mountains, seemed to be turning over into the wind. The Highlands are a succession of natural dramas, and here was a splendid, last-act curtain, as the hills showed their final, stripped-down fangs, leading us to the rim of the earth, with only the empty ocean beyond. We had not seen a person, or a car, or a house, or a tree, for many miles: just rock, moor, brown streams, clouds and sky. We were going. I should think, about sixty miles an hour, and we were exhilarated as we drew towards Cape Wrath.

Then there were two tremendous thumps on the side of the car.

We stopped, and considered, not daring to inspect the damage. But the road was smooth. There were no stones on it. Was it possible that some *loon* had concealed himself, and hurled two gigantic lumps of peat at us passing strangers, with unerring aim? Or had we hit, at long last, the dreaded *gumely jaup*? But there was no sign of spray, muddy or otherwise. Nor did a loon appear, to grin his savage satisfaction.

We got out and looked at the car. Not a scratch. The engine had not fallen out. The wheels were still there. Moreover, the thing worked. So we drove on, mystified.

On the six o'clock news, we heard that Concorde had passed up western Scotland that afternoon.

It is not generally known (that is, George and I did not know) that you cannot get to Cape Wrath by car. The Cape stands at the top of a huge area of mountainous moorland which is completely cut off from the rest of Scotland. The road at the south comes to an end at Sheigra. The road on the east is cut off from the Cape Wrath massif by a long tongue of water called the Kyle of Durness. And at the head of this kyle – 'kyle' means a straight or a narrows – we saw a funny thing. A man was being carried across the water on the back of a woman. The water, I should add, was quite shallow, and not very wide. So the only reason why the woman was carrying the man, so far as I could see, was that he was too lazy to take off his shoes and socks. Once she deposited him on the dry ground, he walked briskly off. Plainly, Women's Lib has not yet reached Cape Wrath.

This business of women carrying men has been going on for a long time. An English visitor to Inverness in 1725 was scandalised by the behaviour of the fishermen's wives: 'The women tuck up their garments to an indecent height and wade to the vessels . . . they take the fishermen on their backs and bring them on shore in the same manner.' Women have had a pretty raw deal in Scottish history. They were the chief victims of Presbyterian morality. Proportionately, many more women were burned as witches in Scotland than in

England (indeed, in the Highlands, witch-burning was an early form of eviction, used against improvident tenants, particularly women). Innumerable local laws were designed to enforce masculine supremacy. Thus, seventeenth century Perthshire laid it down: 'No wife to drink unless her husband is on the premises.' Until well into the eighteenth century, it was very unusual for women to wear shoes, though by that time most of the men had got them. Even when Boswell and Johnson came to Scotland, it caused no surprise if a laird's daughters went barefoot. All the same, it still came as a bit of a shock to see the woman carrying the man.

But George was very taken with the idea. He does not share my progressive notions about equality of the sexes. He was delighted to find himself in a place where women appeared to be treated as beasts of burden; it was, he said, the first sensible Highland custom he had come across. He foolishly supposed that this was the normal, even the sole, form of transport across the Kyle; and when we reached a little pier, on which, admittedly, a group of women were lounging about, he expected them to tuck up their garments to an indecent height, and colport us across.

In fact there is a ferry, a little boat which holds about six people, and is equipped with an outboard motor. The ferryman lives on the other, that is the Cape Wrath, side, in a little house up the hill; and a notice on the pier says: 'To attract the attention of the ferryman, go to a prominent place and wave your arms vigorously.' So we went to the top of a grassy ridge, waved our arms, and the ferryboat came. The kyle looks very narrow; so narrow that you imagine even a weak swimmer could get across without difficulty. But these things are always very deceptive, and in fact it is a dangerous place. The ferryman told me that even a moderate amount of wind could sometimes prevent him getting across in his boat, though it looked solid enough. The kyle is a kind of estuary; the tide goes out leaving only a deep channel, and miles of tempting sand; then it rushes back with great speed and ferocity. Moreover, when it fills up, and when it goes out too, cross-currents and whirlpools are set up. I was surprised to find, crossing this stretch of sea, angry waves and spouts of water suddenly erupt for no discernible reason. And sometimes the incoming tide takes the form of a bore, and thrusts a savage wall of water across the kyle. The ferryman said that he made up to thirty crossings a day in the season, and he must know every inch of the place; but he treated it with respect.

The ferryman's colleague had a minibus; and in this we were taken the fifteen miles or so to the Cape, along an old and twisty track. The countryside, of high moors and ravines, is wholly uninhabited and exceptionally wild, even by Highland standards. There is plenty of wild-life, and for the first time we saw deer at low altitude – a rare sight in June. And of course this is famous bird-watching country. Nearby, in fact, there is the island of Hanna which is entirely devoted to this pursuit. It is uninhabited, but an old bothy has been converted into a bunk-house; for twenty-five bob a night (you take and cook your own food) you can live there and study birds in complete tranquillity and seclusion, all day. I suppose you could camp, too, on the moors near Cape Wrath. There is no one to stop you. But I would not advise it. The whole area has been a shooting-range for many years now, and is alive with unexploded shells.

One of the horrible, if minor, by-products of the arms-race is the way in which such beautiful and desolate areas are seized upon by the military and subjected to grotesque acts of violence. A friend of mine took part, as a fire-control officer, in the final night-shoot held off Cape Wrath by Britain's last battleship, the *Vanguard*, now mercifully reduced to scrap. Of course a warship is a rolling gun-platform, and the range of error is enormous. It is not unusual for a sixteen-inch shell to miss its target by two miles or more. The *Vanguard* fired a broadside straight into the cliffs, bringing tens of thousands of tons of rock crashing into the sea, and destroying the lives and homes of countless sea-birds. The navy's shoots at Cape Wrath have now been cut down to three or four a year. But the RAF, I was told, use the range once a fortnight. They are far worse than the navy for, in addition to the destruction, there is the hateful noise their low-flying aircraft create over a huge area.

It was a very fine and clear day, and we could see not only

Lewis in the Outer Hebrides, but the little uninhabited island of Rona, which is much farther away to the north-west. Of course the track, or road, ends at the lighthouse on Cape Wrath itself; there is only a path, if that, along the cliffs to the south. But I am told – we had not, alas, time to go there – that it is one of the finest stretches of coastline in the world. It includes a vast beach, called Sandwood Bay, into which the waters of an inland loch pour. To get to it, however, unless you are a long-distance walker, you must cross back over the little ferry, and drive an immense distance to Kinlochbervie, a little fishing-port. The road ends four miles beyond, and then you must walk another few miles to the beach. The best things in the Highlands are well guarded by distance.

At Cape Wrath, or rather on the 'civilised' side of the kyle, we stayed at a simple and unpretentious but thoroughly agreeable hotel. There was a wooden drawing-room, with deep, comfortable arm-chairs, and broad views over the water and mountains; and an old-fashioned dining-room. The hotel was full (I warn any reader who wishes to follow in our footsteps that, in the northern Highlands, all accommodation is crowded even early in June, and it is prudent to book ahead), so George slept in a wooden pavilion at the back, and I in a stone cottage; we found hot-water bottles in our beds. At dinner that night, for a modest sum, we had delicious leek soup, freshly-caught salmon-trout, local mutton, and a fine gooseberry pie. This is the kind of food George and I like best: we had a bottle of excellent claret, too. Moreover, there was a bar nearby, where we found some rough and jovial fishermen, and some pretty, giggly girls, who would have been happy, I daresay, to lift their garments to an indecent height and transport the fishermen out to their boats. So altogether, to quote Boswell, 'we were well pleased with our treatment at this place.'

The only little town in the neighbourhood is Durness, which we inspected without much enthusiasm. True, we looked over the famous local cave, called the Smoo Cave, which is a perfectly good cave, as far as caves go, which is not very far in my opinion; and we looked at Lord Reay's house on the beach; and we saw the ruined chapel, which has a pre-Reformation font and the tomb of a sixteenth-century warrior. But the trouble with Durness is that much of it consists of the detritus of military installations; it has a ragged and improvised air. During the war it was an important military base. Round the corner, on Loch Eriboll, the entire German U-boat fleet came in to surrender; while from Durness itself the RAF operated a Coastal Command headquarters. To-day, little or nothing of the Service presence remains, except their concrete and corrugated-iron rubbish, and their mean-looking prefabs. In 1964, the county council bought the main RAF barracks and turned it into an arts-colony. It does not look very pretty, but it seems to function. You can live there all the year round if you want, or stay at its hotel for short periods. The colonists work as individuals, but sell their products through a co-operative scheme. There, we bought some very pretty screen-printed silk scarves. George, in his abrupt way, asked the modest young man who designed and printed them: 'Why on earth did you come here?' The young man replied: 'To practise my craft.' Even George agreed that this was a good and sufficient answer.

CHAPTER FIVE

In the valley of Strath Naver – An encounter with the upper classes – The story of the clearances – Highland pamphlets – A polite small boy – A nonexistent earth-house and a laird's table – Dornoch Cathedral and its monument – The Sutherland family museum – Farewell to the Highlands.

A few miles past Borgie Bridge we turned south off the main road which takes you to Dunnet Head and Wick, and went up Strath Naver, the long and solitary glen leading into the heart of the Sutherland interior. This was the scene of perhaps the greatest crime in Scottish history – a crime in British history, too, for the Westminster parliament did nothing to stop it. The massacre of Glencoe was merely a squalid and brutal episode in the long story of Highland gang-warfare and its suppression; but the Sutherland clearances were an act of tribal genocide, carried out with the full force of the law, and subsequently vindicated by its majesty and authority – a crime whose fruits are still enjoyed by the class of its perpetrators.

Strath Naver is not a spectacular glen at all. If you did not know its history, you would not be inclined to take much notice of it. It is pretty enough, and quiet, but the hills are low, and there is none of the drama and majesty of a great Highland valley. But it is much more prosperous than most glens. We passed all the evidence of high farming: large, well-ordered fields of flourishing crops; herds of fine cattle, vast flocks of sheep, carefully-preserved fishing, the hills open to hunting, at a price. Obviously all is well managed and, no doubt, very profitable. But we saw no people, and no humble homes; just well-maintained farm buildings. All the evidence that a teeming Highland peasantry once lived there has been tidied away into oblivion. We were unable to find a single ruined croft, or even a heap of stones to mark a forgotten habitation of the poor. Every trace of the old peoples of the glen appears to have gone, for ever.

Actually, it is not quite true to say we saw no people. We came across two smart Land-Rovers which were blocking the road, and whose occupants were having a confabulation, no doubt on some sporting matter. They had the *tenue* and equipment of their class and occupation: elaborate fishing-rods, nets, gaffs, baskets, guns, dogs and servile-looking under-lings. They wore the kind of raincoats favoured by colonels at point-to-points, tweed sou'-westers, or whatever they are called, pulled down over red faces bristling with moustaches, and big rubber boots. Rather grudgingly, they got out of our way; but they looked as if they wished to say: 'Look here, don't you know you're on a private road? We'll have no poaching here. What, trespass on our property, would you? Have you got a licence? Where's your permit, eh? Now then, my man, don't argue. Just open the boot of your car. Salmon, eh? You'll get eighteen months for this, by God. Right, Machonochie, nip up the road and ring for Constable Mac-Touchbonnet.' Of course they said nothing of the sort, and I rather wish they had; for George was spoiling for a row with the upper classes – had been, indeed, ever since a similar type of gent, in a brand-new posh car, had bumped into him while trying to drive off the Mull ferry. As a matter of fact, all the man in the posh car had done was to inflict serious injury on his own coachwork, and so infuriate his own bossy wife. But there had been no opportunity for George to give him a piece of his mind. Some ripe expletives, and other cogent observa-tions, had been simmering gently in George's head all over northern Scotland, in the vain hope that at some stage the gent in the posh car would again cross our path. They would have been willingly and gratefully expended on the Land-Rover gang at the slightest hint of upper-class impudence. But it was not to be; and we drove on, George muttering to himself.

No one seems to know exactly how many people were expelled from Strath Naver during the clearances. At the beginning of the nineteenth century, the Countess of Sutherland, whose title and estates descend through the female line, and her husband, the Marquess of Stafford, owned more than two thirds of the county of Sutherland, and were determined to carry through 'improvements'. Between 1807 and 1821 their factors turfed out up to 10,000 tenants-at-will and turned the land over to sheep. The clansmen thought, quite erroneously, that they had an ancient and inalienable right to their land; but in law they had no case whatever. They were true, Gaelic-speaking peasants, miserably poor, who grew just enough food to live on, but who saw their real function as providing fighting-men for their chief: it was reckoned that 25,000 of them in Sutherland could raise 2,000 first-class warriors. But to the utilitarian James Loch, Lord Stafford's commissioner, who devised the plan for the clearances, they were an anachronism. The only way to raise both their living-standards, and the revenue from the land, was to shift them out of the glens and down to the coast, where they could be put to profitable and time-keeping employment. As he wrote:

> Contented with the poorest and most simple fare and, like all mountaineers, accustomed to a roaming, unfettered life which attached them in the strongest manner to the habits and homes of their fathers, they deemed no new comfort worth the possessing which was to be acquired at the price of industry; no improvement worthy of adoption if it was to be obtained at the expense of sacrificing the customs or leaving the hovels of their ancestors . . . did there ever exist more formidable obstacle to the improvement of a people?

Here was the familiar, high-minded ruthlessness which characterised enlightened society in the first stage of the Industrial Revolution. Of course it was hoped that many of the crofters would emigrate to make their fortunes in Canada, the United States and Australia, as some eventually did. Plans were made, however, to create model villages, whose people would work at the textile trade, to build fishing-ports and set up fish-curing plants. A good deal of money was actually spent on such projects, harbours dug, roads and bridges built, and villages and inns brought into existence.

But the positive side of the scheme failed almost completely. There was a slump in the fishing-trade, and nothing came of the textile project. The only part of Loch's ambitious plan to be carried through to the bitter end was the clearances themselves. In January 1814, Patrick Sellar, the chief factor to the Sutherland family, who had rented Strath Naver from them for sheep farming, issued quit notices to the clansmen. In June, when the notices expired, he went in with force. The able-bodied men were up in the hills; only women, children and old people were left to defend themselves. Sellar's men pulled out the roof-timbers of the houses, and set fire to the rest. Some of the tenants were not given time to get out their possessions and animals, and two old people actually died. It was a horrible scene, which aroused great resentment throughout the area. Under popular pressure, the local Sheriff-substitute prosecuted Sellar for arson and culpable homicide. But the defence argued that the trial was 'in substance and in fact a trial of strength between the abettors of anarchy and misrule, and the magistracy as well as the laws of this country'. The court came down heavily on the side of 'law and order'. Sellar was triumphantly acquitted, and the Sheriff was dismissed and forced to apologise to his betters.

What made matters worse was that the kirk sided solidly with the lairds. In its sixteenth-century origins, the Presbyterian church in Scotland was in many ways a progressive one. It successfully strove to give Scotland the best primary and secondary educational system in Europe, indeed in the world; it also crusaded, less forcefully, to create institutions of social welfare. But as the Establishment, and especially after the Union, it identified itself increasingly with the powers that be, and preached the doctrine of absolute obedience to the duly-constituted authorities. That way lay salvation in the next world, if not happiness in this. The ministers of the kirk went through the glens telling the people that, if they resisted the law, they would certainly burn in Hell; and they took the lead in organising and instilling a cap-doffing servility towards the lairds. In any case, the clansmen had no chance to offer successful resistance. The Royal Scots Fusiliers were available to back up the civil arm. Despite their name, this regiment

was recruited mainly from poor Irishmen, who were burning
to revenge the suppression of the Great Rebellion of 1798,
carried out in great part by troops brought from the Highlands. Thus Britain, as often in the past and the future, was
able to use the regimented poor from one dependency to
demolish the rights of the defenceless poor of another. (As,
indeed, to-day, Britain uses Scotch troops to keep in subjec-
tion the Catholic poor of Ulster.)

Opinion about the clearances has swung round completely
in the last century. The full story has been told in John
Prebble's magnificent and fierce book, *The Highland Clearances*,
which everyone who visits this part of the world should buy
and read (it is available in Penguins). At the time, the Gaelic
poets fulminated against the clearers:

> Destruction to the sheep from all corners of Europe! Scab,
> wasting, pining, tumours on the stomach and on the hide!
> Foxes and eagles for the lambs! Nothing more to be seen of
> them but fleshless hides and grey shepherds leaving the
> country without laces in their shoes. I have overlooked
> someone, the Factor! May he be bound by tight thongs,
> wearing nothing but his trousers, and be beaten with rods
> from head to foot. May he be placed on a bed of brambles
> and covered with thistles. Thus may this stray cur be driven
> to Atholl.

But the chief propaganda work on behalf of the dispossessed
was carried out by one Donald Macleod, a stonemason from
Strath Naver, who had been evicted and saw his wife go mad
in consequence. His experience showed the power of the kirk
ministers. It was difficult for a poor man to seek legal redress
unless he had a certificate of good character from the minister;
Donald was refused one 'because I obeyed not those set in
authority over me'. But from Canada, where he emigrated,
this bitter and pertinacious man kept up for nearly forty years
a fusillade of angry pamphlets against the Sutherlands, which
in time transformed the public vision of these events. All the
same, no retribution befell the Sutherland family, who indeed
did nothing illegal; and they still own vast tracts of the
county. The Scots brood constantly over their historic in-

juries, but take no practical steps to efface them; like the Irish, they are fonder of a grievance than of a remedy.

Incidentally, the habit of publishing furious pamphlets, to set right the historical record on ancient injustices, is one of the most pleasing characteristics of the Highlands. Up in Sutherland, I bought an excellent work, called *Three Centuries of Falsehood Exposed: The Shame of* 1650 *laid upon the Highland Laird Neil Macleod, Eleventh Baron of Assynt,* written by the Misses M. L. and E. A. Macleod of the Cadboll branch, Easter Ross, Bee Jay, of Gairloch, Wester Ross, and Colonel Roderick Macleod, DSO, MC. Like most people, I was not aware of the story of Neil Macleod; but I am now convinced he was wrongly judged by his contemporaries and history; and I am delighted that the facts have at last been put straight. May many more such pamphlets be published! They assuage, at trivial expense, the wrath of the writer; and give harmless amusement to the casual reader.

There are more inhabited castles in Scotland than in any other part of the British Isles. I do not begrudge this fact, for, if there are to be lairds at all, it is better to have them, huge, hateful and visible, in a castellated setting of power and arrogance, than tucked away in the Bahamas and other tax havens, raising rents from afar. In a negative sense, inhabited castles keep one's revolutionary spirits up. But of course some of these castles have fallen on evil times, or have been made over, or sold, or let, for other purposes. We prowled around one such which had been turned into a kind of school. George disappeared, and lost himself in the grounds, and after a bit it came on to rain, so I sat in the car. The car was parked near a great door of heavy wood, with elaborate metal fittings, and in due course a small boy appeared, and began energetically to push against it.

From the car, I called out: 'Boy, come here.'

The boy came. He was, I suppose, about ten, and rather on the small side. His blond hair was a little long, but tidily cut; and he wore a neat suit, and a neat collar and tie, and his shoes were smartly brushed. What a well-turned-out boy, I thought, a credit to his establishment, and a standing rebuke to his scruffy contemporaries in England.

'Boy,' I asked, 'do you attend this school?'

'Oh, yes, sir,' said the boy. 'Certainly, sir. This is my school.'

Admirable small boy! How agreeable and how rare, I reflected, to find archaic politeness among the young these days.

'Well, boy,' I said. 'This seems a nice school. Is it a nice school?'

'Oh, no, sir,' said the boy. 'As a matter of fact, I think it's f—— awful.'

And he returned to the door, pushed against it, and went within.

We now drove south, and to some extent east, and met with many misfortunes in our attempts to inspect local antiquities. First of all, there was the case of the non-existent cairn, marking the spot of a *souterrain*, or second-century earth-house, which George was particularly anxious to see. I was less anxious, for the description we had of it said that it was dark and liable to be wet; that one had to crawl into it backwards; and that a torch was advisable. We had no torch. However, we set out to find it. But the cairn seemed to have disappeared. There were, to be sure, many odd-looking rubbishy ruins in the neighbourhood, but as I pointed out to George, there was no guarantee that any of them was of great antiquity. In 1679, one Thomas Kirke described the dwellings of the crofters:

> The houses of the commonalty are very mean, mud-wall and thatch the best: but the poorer sort lives in such miserable hutts as never eye beheld; men, women and children pig together in a poor mouse-hole of mud, heath and some such like matter.

Many of these huts, moreover, had no opening except a hole, through which you had to crawl on hands and knees. How could an ordinary person tell the difference between such places, built in the eighteenth century, and a historical monument from 1500 years before? At last we found a pile of stones which might have been called a cairn, and nearby a low, drystone construction. George tried to get in it, but all it contained was sheep-droppings, a page torn from an old copy of the *Scottish Daily Record*, and an ancient and evil, but unmistakably twentieth-century, boot. George said it was all the fault of my map-reading, which had attracted, from time to time, some of his more pointedly critical observations, and which I felt in all justice, more properly belonged to the gent in the posh car.

We likewise failed to find the important monument to the Great Steam Plough, first used by the Duke of Sutherland to impress upon his tenantry the advantages of modern science, the power of his purse, and the merits of 'improvement'. On the other hand, we found a curious thing not mentioned in any of the guide or local handbooks. This was a tree stump, about five feet high, by the side of the road; and on top of the stump had been firmly inserted an old but still serviceable arm-chair. There was nothing to indicate the purpose of this arrangement. But perhaps, we thought, it was reserved for the local duke, to take a rest in the course of peregrinations through his domain. If so, he would need to be hoisted on to it by several powerful gillies. Local dignitaries in the Highlands do have landmarks called after them. In Skye I remember having seen two big, flat-topped mountains, known as 'Macleod's Tables'; they are well described in that admirable little book, now long out of print, *A Summer in Skye*, by Alexander Smith. There was, we discovered, a similar protuberance in Sutherland, called Lord Reay's Green Table. But when we reached the point indicated on the map, there was nothing remotely resembling such a thing to be seen. Some miles on, however, we found a curious fluke of nature. Some cataclysm, or torrent, had washed away the earth and stones from around a squat pillar of rock, surmounted by a mossy top, about twenty feet across. It was certainly green; from a distance it might have seemed like a table; and it could have been used for Lord Reay's picnics, assuming enough gillies to get him and his guests on to it. Not wishing to pursue the matter further, we concluded that this was indeed Lord Reay's green table, and I did a sketch of it, to be seen overleaf.

Heading now for the North Sea, we passed over a gigantic moor, which stretched for many miles without a solitary sign of habitation. Vast forests have been planted there by the Commission. But the trees are still small, and even these ambitious plantations encompass only a tiny fraction of the moor. It was a desolate and depressing place. The rain streamed down relentlessly. The sky was black, and the clouds pressed down upon the sodden heather and peat-bog, so that the horizon vanished in a grey blur. We were glad to reach Lairg, at the end of the moor, and then travelled across Strath Fleet, to the smart, tidy and historic town of Dornoch.

Dornoch, we found, had been an ancient cathedral city. The cathedral, in fact, is still there, though it is used by the established kirk, and is called a 'historic church'. After the reformation, of course, it fell into ruin, and the Sutherland family built it up again in 1835–7. But, as the official guide sharply points out: 'If the cathedral had the good fortune to be built when ecclesiastical architecture was in the zenith of its glory, it had the misfortune to be restored at a time when it was at its lowest ebb.' Anyway, the church, or cathedral, is now by way of being a monument to the Sutherlands. The wicked Marquess of Stafford, who put through the clearances, was made the first Duke of Sutherland, presumably because he had succeeded in thereby raising his income to ducal status.

When he died, in 1833, the kirk ministers of the county
cajoled or threatened the local people into an enthusiastic
demonstration in his honour. More than 10,000 people, led by
an unctuous clergy, attended his funeral; and afterwards
subscriptions were raised, or exacted, to perpetuate his
memory in physical form. He was buried in the cathedral, and
his marble tomb, erected by the 2nd Duke, dominates one end
of the place; the inscription reads that he died 'amidst the
universal grief of his Scottish and English tenantry'.

This was bad enough; but worse was to come. Not only were
further memorials erected by grateful tenants in Staffordshire
and Shropshire, but not far from Dornoch, on the top of a
high ben overlooking the sea, a 90-foot monument was put up,
surmounted by a giant statue of the duke, carved by Sir
Francis Chantrey. This outrageous affront to the dispossessed
of Sutherland is still there; it is inescapable; it dominates a
vast area. Why haven't the Scotch found the courage to
tumble it down Ben Bhraggie (that well-named hill), as the
Hungarians overthrew the colossal bronze statue of Stalin in
Budapest?

Sutherland wealth built their great castle at Dunrobin, in the
shadow of the statue. It is Victorian baronial, for the most
part, not unlike the extraordinary palaces built by mad King
Ludwig of Bavaria; but much of the old castle remains, and as
you walk along the terrace, and each new facade springs into
view, you descend backwards in time, to the end of the Middle
Ages. I strongly recommend a visit. The gardens are superb,
and well-tended; there is a fine high fountain; and best of all,
a family museum. This museum is nothing more pretentious
than a collection of mementoes and trophies. But, happily, it
has not been modernised and invested with all the apparatus
and clinical tidiness of curatorial scholarship. It is thus a
perfect specimen of a pre-1914 country-house museum, with
some between-the-wars additions. There you can see a giant
alligator shot by Lord Stafford in Ceylon in 1908; an Indian
gawr shot by the Duke (crossed out) the Duchess of Sutherland
at Mysore in 1930; the head and neck of a giraffe, and many
other murdered and stuffed animals. These Sutherlands were

evidently great slaughterers of beasts, as well as drivers of crofters. But there are many other quaint survivals: an old Baby Daisy vacuum-cleaner, an antique spin-drier, heavy-looking implements, and archeological bits and pieces, together with the funny white hat worn by Lord Raglan when he was directing his disastrous operations in the Crimea. There is even the invitation the family received to attend the coronation of Edward VII, which is hardly an invitation at all, more a rude threat: 'These are to will and require you and the Duchess your wife (all expenses set apart) to make your personal attendance.' But the majestic phase of the family history appears to be drawing to a close. The present duke is expected to be the last. The castle has been inherited by a woman, and the estates are not what they were.

'Such are the things,' wrote Dr Johnson in the last paragraph of his book, 'which this journey has given me an opportunity of seeing, and such are the reflections which that sight has raised.' A Tour, concluded Mr Boswell, on which 'I shall ever reflect with great pleasure'. I would add *amen* to both these remarks. We had travelled 2,000 miles or more, by my computation, seen many interesting and remarkable things, and met nothing but courtesy from the ordinary people of the Highlands. If anyone who has not yet visited this splendid part of Britain should be induced to do so by the record of our jaunt, we shall be happy to have been of service, both to them and to its inhabitants. We are grateful to our two mentors, whose journey almost exactly two hundred years ago inspired our project; and we salute their illustrious shades.

One final point. Long before we set out, George had broken the key of his car in the ignition, and had lost the key of the boot. So we had no means of locking up our possessions and indeed took no steps to protect them. But though they were constantly left unattended, nothing was ever touched; and we concluded that the Highlanders, whatever other merits they might or might not possess, were certainly honest. At Perth, we took the car-sleeper to London; and during the early hours, while the train was shunting in the outer fringes of the metropolis, thieves or hooligans boarded it, and robbed two cars. We had nothing stolen; but a window of George's car was smashed, and in the morning we found its interior littered with thousands of glass fragments. 'At least,' said George, 'we know we are back in civilisation.'

PART TWO

George Gale's Narrative

CHAPTER ONE

*Up the M1 and M6 to Ullswater – Gretna – Glasgow's
Glaswegianality – Loch Lomond – Inveraray, town
and castle – the Argylls – Portsonachan –
growing rabbits*

Paul had suggested to me that he and I should do a tour of the
Western Highlands and Islands of Scotland, bearing in mind
the celebrated journey to the Hebrides made by Samuel
Johnson and James Boswell in 1773. What Paul's exact
motive was, I do not know. Mine was to see this north-west
land and sea, to which, for as long as I can remember, I had
been drawn. We had a secondary motive in common: to write
an account of the jaunt.

Paul is one of the lean and hungry kind with bright and
orange hair: he takes very unkindly any suggestion that he
might dye it, as once I learned when I remarked that his hair
seemed to grow brighter over the years whereas mine grew
steadily dingier. He used to be very excitable: one of the
angriest of young men, with a nice capacity for, coining the
word, *ruderies*. He has not entirely lost this capacity: I recall a
splendid breakfast a couple of years ago that was dominated
by his noisy and urgent requirement that he be supplied with
the correct silver teaspoon with which he was accustomed to
eat his boiled egg. He is a socialist, a Roman Catholic and a
Christian, and I am none of these things, so that our dis-
agreements have always been conducted on a civilised level.
He goes for walks with no other purpose than to walk; and
when he walks he strides, and takes a stick, what's more, and,
if available, a dog. He goes to bed early and rises early. He is a
very vigorous chap, full of vigorous opinions and gestures. One
of his most engaging characteristics is the vehement certainty
with which he asserts the great bulk of his anecdotage and the

whole sum of his ideological equipage. This makes him a formidable journalist and controversialist.

He is one of the few people I know who are both fun to be with and funny to contemplate. Although he is a natural autocrat, he is aware of his own solemnities. He knows many more historical facts than I do, although it is my conceit that his understanding of historical processes is less assured than mine. He is Oxford, French-speaking and bibliophilic, all of which may be considered suspicious. On the other hand, he knows and enjoys English water-colourists and thinks that ballet is rubbish; and we are also agreed that the Common Market venture is both foolish and a betrayal. Essentially, he is a man whose instincts are much sounder than his convictions; and this, together with his more than somewhat liverish sense of humour, makes for a good friend and a good companion. He made a fine navigator throughout the jaunt during which I was the driver.

Thus, on a beautiful afternoon in early June – it was, in fact, Derby Day – having left behind the cares of offices and homes, we drove north up the M1, past Birmingham, out on to the M6, crossing over the Manchester Ship Canal, leaving Preston and then Lancaster behind. By Sedbergh we passed some moors which Paul recalled from his schooldays, we climbed Shap by the new road, and then we saw the Lake District hills ahead and the Pennines' highest point, Cross Fell, to the east.

'Look! There's Saddleback!' said Paul, pointing out Blencathra to our north-west.

'I feel myself to have been getting happier the farther north we are getting,' I said: and it was true.

After crossing over the Ship Canal the red sandstone started showing through the threadbare soil of the northern English moors. We both knew and loved the Lake District; I had spent the war years at school there. We left the motorway at Penrith and took the road to Ullswater, where we had decided to spend the night. At Yanwath we stopped for a drink at a pub called Yanwath Gate, which was the first pub I had ever been in, almost thirty years before, when it was called the Yanwath Yat. Then, it had been dark even in the brightest day, with old men crouched beside the smoky inglenook playing dominoes and a woman who shuffled into the back room to pour their beer and our cider out of barrels. Now, it had all been brightened up, with Spanish-style decorations, and the beer was keg. The pub had changed more between my last two visits than ever before, in its two or three or four hundred years. Paul arranged for us to stay at a hotel on Ullswater and the pleasant young woman behind the bar wouldn't take any money for allowing him to use her telephone.

At the hotel the dinner was adequate and after it Paul and I walked down to the lakeside through a meadow high with buttercups and higher with midges.

'The next-door hotel would have been better,' I said.

'This is perfectly all right,' said Paul.

We came to the lake. There was a derelict boathouse: 'That would make a better house for a man than for a boat,' I said, loudly. We rounded the side of the boathouse and a man and a boy, quietly fishing, were both startled and angered at us intruding on them. The lake was very quiet, and I shut up. Across the lake is Gowbarrow Fell: I remembered its line, and other skylines to the west and north; but not the skylines to the east, save the Beacon, the hill above Penrith, and the arched back of Cross Fell. We walked a little farther on, where a perfect bay was formed within the perfect lake. All along the bay were yachts and boats and floating caravans.

'They spoil the thing,' I said.

'Of course they do,' said Paul.

'I'd like a boat somewhere here. With a boat it would be easy. You'd always get the view.'

'Yes.'

Back at the hotel we had a couple of whiskies and Paul retired, having turned down abruptly a suggestion that we might go off to a pub at Tirril, where I first remember having thought of a girl as a girl, at a dance. A rugby-playing prefect later got her in the family way, as we thought of it, jealously. Up in my room I looked through the window at the lake and the hills, navy blue and grey, and the very cold white moon, felt refreshed, and went up to bed and to sleep.

There was much birdsong in the morning, which normally I

dislike; but at quarter to seven I looked again out of the window, on to the garden, lake and hills, and again found pleasure and solace in the view. The rhododendrons were fat and full, a crow cawed, the lake had a stillness of flat water that I had forgotten about, and the yachts, motorboats and waterborne caravans were tolerable.

At Penrith I bought a pair of shoes. The shop woman talked about how they would repair them: 'They should be whole-soled,' she explained, tapping the leather, 'don't have them half-soled. We'll always do it for you. Are you sure they fit? Are you sure?'

Back in the car, Paul said: 'They talk a lot up here.' Then, with Carlisle behind us he said: 'I wonder if they'll tell us when we're in Scotland?'

'I haven't the faintest idea,' I said, 'but I remember, before the war, driving with my parents from Newcastle into Scotland by the East route, or perhaps by Carter Bar. After we'd crossed the border there was a man dressed up in a kilt playing the bagpipes. I suppose he was begging. He would pose for pictures to be taken of him. Like Egyptians do with camels at the side of the Sphinx.'

We crossed a temporary bridge. 'Scotland,' said a sign on our left. 'The first house in Scotland,' said a sign on the right; '10,000 marriages celebrated here.' This was Gretna.

Soon Paul said: 'It looks poorer already,' and his observation pleased him. His pleasure displeased me, not because of the inaccuracy of the remark but because I took it to mean that he was designating himself to be the newest and latest hammer of the Scots. Paul was very heavily armoured in Johnsonian quotes and also, having just completed his *The Offshore Islanders*, in English history. He was to be Johnson, which in name he already was, and was leaving me to play Boswell, I supposed.

I am not particularly fond of Dr Johnson: it needs to be explained to me, from time to time, why he is as great as he is generally presumed to be. He neither irritates nor angers me enough for me to deny his greatness: I concede it, and put him in the class of worthy bores like Sir Walter Scott and, alas, Dickens. If Johnson actually talked as Boswell reported him, then he was rude to his intellectual inferiors unless they were his social superiors, pompous, filled with his wind and short on wit. He was a 'character' which, since television, we call a 'personality': like Gilbert Harding, except that the comparison flatters Harding and insults Johnson.

Sam Johnson was not, of course, a serious, world-class bore like Marcel Proust or Charlie Chaplin. He was amusing and intelligent on many occasions. He did not produce rubbish. His Dictionary was, is, magnificent. I have seen his house, and eaten at the *Cheshire Cheese* and have, like him, made a living scribbling around Fleet, or Grub, Street. Kingsley Amis tells me the *Lives of the Poets* is first-class. He ought to be more of a hero to me than he is: it may be a matter of tone-deafness; or that he was Tory, not Whig. He, even by the standards of that stinking age of elegance in which he lived, must have noticeably stunk. Fat men stink more than thin ones, I believe; certainly fat women stink more than thin ones more often than not, and when they don't, they look as if they did, which is as bad. There are more nasty smells in the world than nice ones: we would be better off without the sense, or I would, who am asthmaticky and hay-fevered. Dr Johnson must have belched and farted a lot and have been noisy and dribbly in eating his food and slopping his wine. Proust and Chaplin must surely have been tidy feeders; not the people to have to dinner. Dr Johnson would, all in all, be a most welcome guest. Being also a celebrity, he would be doubly welcomed, for celebrities are people with whom one puts up, and sometimes puts up.

James Boswell must have been a more engaging fellow: better to have been him than Johnson, a womaniser rather than a fat white-skinned soft virgin, in no way a bully or a boor or a bore. Still, he certainly played second fiddle to Johnson; and so I thought that Paul, in saying, just north of Gretna Green, 'It looks poorer already,' was establishing himself as first fiddle. I replied, 'Oh, balls.'

Soon the scenery became Scottish: we passed a church, the stones white-painted except for the larger and better-trimmed corner-stones which were painted black. Churches like these, severe, melodramatic and smug, are to be found all over Scotland, Wales and Ireland and in the western edges of England. Are they what being Celtic is about? I was brought up to think that I was part-Scottish, which was to say part-Celt, and also that it was a good thing to have Scottish or Celtic blood. Celtic places are beautiful but Celtic buildings are ugly. Did the Celts (pretending for the moment that they *were* all Celts) make their landscapes beautiful, or flee to places which happened to be beautiful because they were poor, or seek these areas where the rock breaks through the soil and meets the sea? The church and other ugly buildings are in part the consequence of their poverty; but in part, also, the consequence of that terrible creed which Calvin thought of and John Knox brought, and which infected the poor, weakened so that they had little resistance, like the plague. The poverty is Celtic, the beauty of place is the same as the poverty of the land, and the ugliness is Calvin's and Knox's.

We drove partly around and partly above Glasgow: the huge motorways through the city and across the Clyde were not then quite completed.

'This,' I said to Paul, of Glasgow, 'is the most foreign town in Britain.'

'Yes, and they clear the slums and put motorways in their places.'

'The tenements of the Gorbals are fine buildings. They look ugly to us because we think of the Gorbals as being ugly. Glasgow's slums are fine architecture. Its council is corrupt. Look at this motorway!'

'Look,' said Paul, 'we're in the middle of the lunch break. Look at all those shabby works buses.'

'Those are not works buses. Those are Glasgow Corporation buses. And it is dinner, not lunch.'

Scottish national newspapers are parochial in the way that English local newspapers are parochial. Scotland possesses a pride in itself which constantly requires replenishment. Scotland, too, is aggressive about its Scottishness in the way that little men, when drunk, sometimes seek fights with bigger and stronger and sometimes sober men. The Glaswegianality of Glasgow is different: Glasgow is a big fellow who has made himself ugly; a handsome man of almost classical features, disfigured by an inherent wildness, a propensity for violence and disaster.

Glasgow is a city built in stone; and although it may be difficult to build modest domestic apartments of stone, it is almost impossible to build ugly buildings in dressed and unpainted red sandstone. Glasgow is built of such stone, which seems to enforce the dignity of decent proportions on the windows and doorways of the great tenement blocks. Their tiled passageways are different. The ugliness or beauty of anything depends partly on what it is and partly on what it looks like: it is what they are much more than what they look like which makes the ugly parts of Glasgow ugly.

It is ugly by repute; nor is there anything new in this, for Boswell, in Glasgow with Johnson, who had 'viewed this beautiful city', tells how Johnson earlier and in London had replied to Adam Smith's boasting of Glasgow with: 'Pray, sir, have you seen Brentford?' Boswell reminded Johnson of this 'while he expressed his admiration of the elegant buildings, and whispered to him, "Don't you feel some remorse?" '

When Boswell and Johnson were there, the linen manufacturers and great tobacco merchants had made Glasgow almost as large as Edinburgh; and the industrial revolution, about to break in Scotland as in England, would soon make Glasgow a centre of the cotton industry and the biggest and most powerful town in Scotland. But its slums then, as now, were rough, rougher than any others in Britain. Violence, including racial and sectarian violence, came to Glasgow side by side with industry. The spinners, iron workers and miners of the Glasgow area defended themselves with early trade unions. The textile manufacturers, ironmasters and coal owners often imported still cheaper Irish and Highland labour; and ever since, Glasgow has been the fiercest and most violent (and most revolutionary) of British cities.

They drink heavily and fast in Glasgow, conscious that it is never far from time being called; and nastily, mixing lemon juice with whisky, and pouring down their 'heavy' beer or their draught lager on top of the whisky. Belfast is Glasgow's daughter, or twin, city; and Belfast apart, no other city in the

kingdom has so much the feel of vehemence. It is the most
foreign, and potentially the most frightening (which may be
saying the same thing). There are plenty of places in Glasgow
where it would not strike me as ridiculous to be given the
advice the metropolitan magazine *New York* gave to its
readers—'Walk along the curb of a sidewalk and avoid
shadowy doorways or building recesses . . . If you think you're
being followed and your building is not served by a doorman,
keep on walking . . . Have your keys ready when you enter
your building . . . Organise tenant or street associations . . . If
you are mugged, don't resist . . . Don't hesitate to help someone in distress.' I do not mean to say that Glasgow *is* a
dangerous place. I mean that it is a place I could imagine
becoming dangerous, like the Chicago I imagine, or the New
York which I know and liked. It is Glasgow's capacity to
produce such imaginings that makes it the most foreign of
the towns I feel not lost and abroad in.

And it is exciting, as Edinburgh will never be. Edinburgh is
feminine and Glasgow masculine. Edinburgh is prim and
prissy and pretty and it was praised to me so much, when I
was a child, for its Princes Street, its Arthur's Seat, its
Royal Mile, and its stature as the Athens of the North, that I
never liked it much. Edinburgh possesses that awful assurance
of moral superiority and certainty of rectitude with which I
associate Scottish Presbyterianism; as a result of my upbringing and having been educated into respecting it as a spiritual
home I have ended up detesting it as a sink of hypocrisy.
Glasgow is different: coarser, cruder, stronger, truer; but also,
I regret to say, in the last resort, more sentimental.

We stopped on the roadside above the new bridge across the
neck of the Clyde, then shortly to be opened. Paul drew a
sketch and I took some photographs.

'Let us lunch at the Lomond Castle hotel,' I suggested.

'What is it like?' asked Paul.

'Very flashy position on the banks of Loch Lomond. We
used to drive out from Glasgow to have lunch there.'

'Hmm,' said Paul, suspiciously. 'All right. We will look at
it.'

'We can always have a drink in the bar first.'

We drove up the drive of the hotel: it looked too neat, the lawns trimmed too tidily, too expensive, that is.

'It looks the sort of place we might not get a drink unless we went on to eat,' said Paul.

There were some men drinking on one of the lawns, Glasgow businessmen who had had it made.

'I don't like those fellows drinking on the lawn like that,' said Paul.

'It has some sociological significance, hasn't it?' I said defensively.

Neither of us were on expenses, so we turned back and went instead to the Duck Bay Marina (as I think it was called): a classless place of airport architecture, a big bar, good fast service, no bloody nonsense apart from the décor, and Granadaland food.

'A pint and a half of McEwen's heavy.' Paul telephoned.

'Another pint and a half of heavy.' Paul returned.

Loch Lomond is soft and lovely, but however much one may look at it and admire it technically as a view, the name and the song and the coachloads of trippers and its own conspiracy with the song and the trippers, make it a constant embarrassment. Despite the daffodils, Ullswater, which is not dissimilar, survives much better. The sun was shining bright on Loch Lomond, but we were not sorry to see the back of the loch and to set out for Inveraray and the west: into the Highlands and islands proper.

We suffered a final interruption. Where the road went high above a river a car had crashed through fencing and tumbled down. I thought it had happened much earlier but Paul said it had just happened. We pulled into a convenient lay-by at the same time as another car, coming up the hill towards us. A man got out of this car and said:

'We were picnicking at the bottom of the road. I've come to get some cine-pictures.' Off he went, scrambling down the hill-side.

'Shall we go and look?' I asked Paul.

'No, I don't think so,' he said, 'no one has been hurt. There is nothing we can do.'

I had not thought of doing anything, only looking. One of the chaps from the crashed car was gathering wood and stones to put beneath its wheels and prevent it from rolling farther down. There seemed little point in this: the car was a write-off, and irretrievable anyway.

'They're bloody lucky fellows, those two,' said Paul; and in one way they were, and in another way, which may well have been the way they were considering their situation, they weren't.

Inveraray surprises and delights, both castle and town: historically, a fortunate act of timing, both built when nothing much bad was built; and historically also, an act of town-planning and building. Towns were planted in Scotland to tame and to contain the place as well as to enrich it; as trees and strong grasses are sometimes planted to prevent the soil slipping, or being blown away.

But Inveraray on our visit appeared to have little purpose left but to sustain visitors. I wanted to buy some socks; it turned out to be impossible in the shops of the townlet to buy anything but tartan socks, usually accompanied by matching tartan ties: souvenir socks in souvenir shops. The Castle's fine rooms were spoiled by the crowds and the ropes, necessary to protect and to guide. The rooms, like the shops, contained also a good deal of rubbish, verbal and otherwise:

'You'll have heard of Rob Roy?' said the guide.

'Yes, we've heard of Rob Roy,' the crowd chorused back.

'Well, in this glass case is Rob Roy's very own dirk!'

'Oh!'

'And in this glass case is Rob Roy's very own sporran!'

'Ah!'

Rob Roy is a characteristic folk hero: a brigand and robber of whom little is known. He is thus like Robin Hood, and the various wild men of the American west.

According to the guidebook on sale at Inveraray Castle, the sporran and the dirk handle in the showcases are said to be Rob Roy's because they 'were found on the site of his house in Glen Shira.' (It would be much the same if, out in the wild west, they venerated an old saddlebag found in an abandoned

shack and called it 'the saddlebag of Billy the Kid'; or, come
to that, if the Venetians, supposing them to be short of a
tourist attraction and discovering an old set of best English
sheaths, had called them Casanova's.) However, it is beyond
question that the crowd being shown around the Castle when
we were there were more intrigued and pleased by 'Rob Roy's
sporran' than by anything else in the Castle. The supposed, or
asserted, connection of any relic with some real and famous
person gives the relic great interest; as the Christian church
throughout its history testifies, to its great profit.

Paul said, more than once, during the day we visited
Inveraray, of his Roman Catholicism, 'I am not a convert,
George.' He greatly admired the fine pulpit in the church.

'The main thing in the church is the pulpit,' he pronounced,
adding again that he was no convert.

We bought a couple of hamburgers from a stall, which were
uneatable and which we flung over the sea-wall for the circling
and crying gannets. We next bought two home-made pork
pies, and they were dreadful and we put them in a litter-bin.
The fact was that the crowds of visitors had suffocated this
elegant miniature town, so that it was in the end, as a vital
place, destroyed by those who had come to look and enjoy
and admire. Still, as buildings, as a piece of planning, the town
and castle still owned an elegance which was lifted straight
out of Scotland's best time, the three generations or so between the birth of David Hume (1711) and the death of Sir
Walter Scott (1832). It was in the middle of what the modern
historian of Scotland, T. C. Smout, describes as 'The Golden
Age of Scottish Culture' that Inveraray was built, and Sam
Johnson made his journey to the Highlands and the Hebrides.

I had expected, for no good reason, that the Castle at
Inveraray would be tumbledown, rude, part ruinous. Instead,
I found it to be austere outside, sharp and clean-edged,
without anything ruinous or fungoid or even much weathered,
built of slate quarried on Loch Fyne. Castle and town were
planned by Roger Morris; William Adam, father of John,
Robert and James, superintended the beginning of the works;
John Adam succeeded his father; and then Robert Mylne took
over, completing the town, and the Castle's interior, to the

designs left by John Adam when he left to join his brother,
Robert, in London. It is little wonder that with such a
genealogy, the castle and the town at Inveraray possess such
distinction.

We drove into the place over the confident high arch of
John Adam's Aray bridge. Within, the castle was no castle but
a richly gilded and caparisoned mansion in the Georgian
Gothic taste; and all my notions that the Dukes of Argyll had
long since become impoverished men without any substance
were at once abandoned. I have since gathered that the present
Duke has spent a great deal of money in restoring his beautiful
house and making it fit to be trampled over by the public. The
main fault with its interior, apart from the excessive number
of people gawping at it, is the decorative use of muskets, axes,
halberds, pikes and such like in what is called the Armoury
Hall. This preposterous decoration was completed in 1952 after
a fire of 1877. Another fire, if limited to the Armoury Hall,
would not come amiss. The house, like the town, now suffers
from its museum-like character; but it is not difficult to
imagine it without showcases, without the Armoury Hall; but
with a wild party of fiery men and beautiful women raging
through the cool and restrained opulence of its eighteenth-
century public rooms.

And calmer occasions, too, we know took place, for in the
drawing-room hung with Beauvais tapestries the guide recalls
that 'Dr Johnson and Boswell were entertained by the 5th
Duke and Duchess in 1773, when the Doctor remarked on

"the grandeur of this princely seat" ', and it must indeed have been grand – although the final decorations of the public rooms were not completed till the 1780s.

Something of the wilder and sterner history of Scotland is to be gleaned from the total titles of Ian, present and eleventh Duke of Argyll: Lord of Inveraray, Marquess of the Great Seal of Scotland, Admiral of the Western Coasts and Isles, Hereditary Sheriff of Argyll and Keeper of the Royal Castles of Dunoon, Dunstaffnage, Tarbert and Carrick. It is comforting to recall the modern instances that the possessor of such grandiloquent titles was also the husband in one of Scotland's most celebrated divorce cases, and that the journalist Lady Jean Campbell, his daughter (and Lord Beaverbrook's grand-daughter) underwent a stormy courtship and marriage with Norman Mailer. Although the Campbells of Argyll may never have been as wild as, say, the Delavals of Northumberland, their blood-line runs straight back to the thirteenth century and all that time they have held to the land now known as Argyll; and for much of that time they were fighting. The line quite probably goes back farther, perhaps to Somerled, King of the Western Isles, who drove the Norsemen out of the Hebrides; and in Somerled's blood, half Norse but half British, may well have been the blood of the *Scotii*, the Celts who came from Ireland when the Romans were pulling out of England and formed the kingdom called Dalriada in what is now Argyll.

We drove from Inveraray to Portsonachan on South Lochaweside: it was a lovely long evening, the sky without clouds, and Loch Awe itself not in the least intimidating.

We dined at the pleasant hotel at Portsonachan, where they produced far too fancy a menu: *tomato soup à l'orange*, which was either tomato soup tasting of oranges or orange soup tasting of tomatoes and in any event a culinary effort at once so fruitful and fruitless that it put me off the rest of my meal. The overdone steak was supposed to be rare. The owner came and stood over us and asked us how it was.

'Fine,' said Paul.

I, a bloody coward, said nothing.

Paul was, is, a hungry chap, not exactly continually ravenous like a dog, but hungry nevertheless.

'I am not as hungry as you are,' I said.

'Hmm,' says Paul, his 'Hmm' conveying the moral superiority of a hungry man.

He drank half a bottle of claret and I had a large glass of Glen Morangie—one of several glasses I had that night of this excellent malt whisky which unfortunately is awkward to obtain in England. Back in the bar a man told me that Johnson and Boswell had spent the night in this place. Such legends are ineradicable. This one stems from the following passage of Boswell: 'We crossed in a ferry-boat a pretty wide lake, and on the farther side of it, close to the shore, found a hut for our inn. We were much wet. I changed my clothes in part, and was at pains to get myself well dried. Dr Johnson resolutely kept on all his clothes, wet as they were, letting them steam before the smoky turf fire. I thought him in the wrong; but his firmness was, perhaps, a species of heroism.

'I remember but little of our conversation. I mentioned Shenstone's saying of Pope, that he had the art of condensing sense more than anybody. Dr Johnson said, "It is not true, sir. There is more sense in a line of Cowley than in a page (or a sentence, or ten lines, – I am not quite certain of the very phrase) of Pope." He maintained that Archibald, Duke of Argyle, was a narrow man. I wondered at this; and observed, that his building so great a house at Inveraray was not like a narrow man. "Sir, (said he,) when a narrow man has resolved to build a house, he builds it like another man. But Archibald, Duke of Argyle, was narrow in his ordinary expences, in his quotidian expences." '

'The distinction is very just. It is in the ordinary expences of life that a man's liberality or narrowness is to be discovered,' notes Boswell.

It is amazing to think of that kind of conversation taking place between men drying out in front of a peat fire, the hut filled with steam and smoke. They went on after dinner, not staying the night; and I doubt whether the 'hut' was the hotel;

although this was where they crossed, for this was where the ferry came from. They were heading towards Inveraray at this point, so it may be that Dr Johnson's churlish remarks about the Duke's narrowness were reconsidered later, after his ducal hospitality. On the other hand, it is certainly my experience that great and wealthy men who build themselves palaces, or yachts, or who otherwise demonstrate their riches, are narrow in their ordinary or quotidian expenses.

It was pleasant, if idle, to imagine Johnson and Boswell at the bar, although there had been no bar then. There was no ferry now, though the stone slipway stood, and the stumps of old jetties, bleached, beautiful, old, rotten and useless, stuck out of the loch. I remembered coming to Loch Awe as a child, and being frightened of the place and its name and how wet and misty and windy and grey it had been. This was surely not the same loch – but Paul found a nineteenth-century map which showed a drawing of a fierce and angry neck of the loch as I recalled it. The map was a reassurance. Loch Awe was once Laughaw, later Loch Ow: it means the dark loch, which was more or less as I had thought of it, rather than this beautiful and almost sweet and gentle spot.

Paul went to bed. Behind, in the bar, the conversation turned to rabbits.

'I could never eat a rabbit again; not after the myxy,' said an English farmer, holidaying in Portsonachan, for the fishing, I believe.

'I like rabbit,' I said, for want of anything better.

'We have now got about sixty cats to keep the rabbits down,' said the landlord, 'and we now have dogs to keep down the cats.'

'If rabbits grow like rabbits here, would it not be better for this part of Scotland to concentrate on growing rabbits? Nothing else grows. The rabbits, you could sell them, "guaranteed free of myxomatosis".'

Around the bar they looked at me with some distaste.

'Prime Scotch Argyllshire Rabbits,' I suggested, 'like your Prime Scotch Aberdeen Angus Beef. You could have Argyll Rabbit Houses like the Angus Steak Houses.'

They were not impressed.

'I took two calves to market yesterday,' the landlord said, 'but the price was wrong, so that's how you had veal.'

'I don't like veal,' I said. 'Gluey French stuff.'

'Then you haven't had ours.'

'All veal is gluey, except when it's dry.'

'What we get wrong,' said the landlord, 'is our bacon. We overcook it.'

'Here, you overcook everything except the bacon. Bacon should be overcooked and steak should be undercooked. You have things arse upwards.'

'We have very good eggs,' said the landlord.

'What about venison, then?'

'A stag jumped into the field at the back last night and started the dogs howling. We can't have venison here. It all belongs to Major Fergusson [or some such name, I forget which] and it is sold under contract to Germany. The Germans pay a fantastic price for venison.'

'That leaves you with the rabbits,' I said.

We had another glass of Glen Morangie each. The telephone rang. A woman guest was fetched to answer it. The landlord returned and told a man he'd become a grandfather. The guest returned, almost in tears, and was given a brandy.

'It was ten o'clock this evening,' she said, 'and I know her name already. Clair Ellen. Clair Ellen Webb. Six and a half pounds she weighed.'

'I feel quite weak,' said her husband, the grandfather.

I went to bed, stumbling around the room which Paul and I were sharing (for he had inadvertently – carelessly – booked us a twin-bedded room instead of the two single rooms we could have had at the same price). He momentarily awoke with an abrupt start and grunt, then turned over and resumed sleeping with innocent, almost impertinent silence. I scribbled a while, and waited for a sleeping pill to combine gently with the Glen Morangie.

The grandparental lot passed outside the room some time later, laughing. They had been also celebrating their wedding anniversary.

'It took them eleven years to produce our granddaughter,' the woman guest said.

'Mind you, Ruby, it took you as long to produce our Ellen.'

'I had my five in nine years,' the voice of the landlord said.

'Good batting,' said another guest.

'Good wicket-keeping,' said a fourth.

I went to pee in the gents because, sharing a room with Paul, I could scarcely pee in the washbasin. The gents had a lovely view of the silent lake, with patches of water rippled by fits of wind glittering in the moonlight.

Back in the bedroom, Paul was dreaming, twitching and grunting like Janet, the Alsatian bitch at home, when she sleeps and dreams of God knows what.

'Hmm,' says Paul suddenly, in his sleep.

CHAPTER TWO

*Loch Awe and South Lochaweside – an unlicensed
hotel in Mull – Scottish food – across from Fionnphort
to Iona – an account of Jacob Lipchitz's statue of the
Virgin Mary called* The Descent of the Spirit *in the
cloisters of the Abbey – 'Random Thoughts' – Fergus
Cashin and a dead Pope – a consideration of the
Scots – the question of pilgrimages and retreats – St
Columba and his present successors – on islands.*

It was the Pass of Brander up the neck of Loch Awe which I
had found frightening. This is where the Glasgow-Oban road
and railway line run: down the loch sides there are only
narrow roads, not overgrown but with enough overhanging
leaves and crowded banks and enough grass growing between
the tracks to see how easily the roads could become overgrown and disappear entirely, as indeed have most of the
people.

In the tribal, celtic days of Dalriada there were many people
in these parts: the tribes were partly nomadic, grazing the land
then moving on; but returning. All the land remained under
tribal ownership. and by one reckoning a 'merkland' – 104
acres – was supposed to support 'twelve milch cows, ten yeld
cows, twelve two yr. olds, twelve one yr. olds, eight horses, a
hundred sheep and eighty goats'.

Be that as it may, there was much more life in Dalriada
than there is in Argyll. The cereal grown was oats; the sea was
fished for herring. But then there was much more life in
eighteenth-century Argyll than there is now. In the parishes of
Kilchrenan and Dalavich, in which most of Lochaweside lies,
the population may have risen by the 1790s to 2,000: they
were mainly crofters, but there were woodmen and dykers,
weavers, shoemakers, smiths, taylors, carpenters, millers,

travelling salesmen, a shopkeeper, two innkeepers and five itinerant whisky sellers: the place was self-contained, self-supporting, and paid rents, either in kind from the crofts or in cash from the farms which were being made out of aggregating crofts. And as the sheep farms grew, the people declined: in 1831 the population was 1466, in 1841 943, in 1851 776, in 1891 415. It may be about the same to-day.

The land now belongs to the Forestry Commission or is game 'forests' bare of trees. Kipling is quoted in a useful little *History of South Lochaweside* (in which I am delving):

> *They shut the road through the woods*
> *Seventy years ago.*
> *Weather and rain have undone it again,*
> *And now you would never know*
> *There was once a road through the woods*
> *Before they planted the trees.*

The old road is still to be seen above the present one, along the loch side; and 'An old man, alive at the start of the present century, could remember counting the lights of six illegal stills among the trees, between Portsonachan and Balliemeanoch. There is said to be whisky still lying in a bog, hidden from the exciseman!' There were four schools in the parish of Kilchrenan in 1792, and a ruined schoolhouse stands by the old road. The last school, in Cladich, closed in 1969.

This is old land: there were neolithic people living hereabouts two or three thousand years before the Celts first came from the east. There are burial places high up above the screes, and in the lake-isles.

From this small, amateur and beguiling *History of South Lochaweside* I take two legends, lacking poetry but giving the flavour of the place, which is very much a highland flavour, as it was until recently (and would be still, were there enough people in the Highlands apart from the tourists, the game people, the forestry people and the hydro people).

There is a legend of the Big Beast of Loch Awe: 'Some say he was like a horse, others like a great eel. He had twelve legs, and when the bays of the loch were frozen in winter, cracks and rumblings were sometimes heard, and the people, with a shiver, would say, one to another: "That is the big beast breaking the ice." '

And there is the legend, or Celtic tale, of Innish Errich, the burial island:

'Armar and Daura were betrothed. Daura waited in her Father's house for Armar to come and claim his bride. Erreth, seeking vengeance for the killing of his brother by Armar, lured Daura with the tale that Armar was lying, seriously wounded by a stag. He, Erreth, had flayed the slain stag and wrapped Armar in it, to keep him from freezing to death, and had left him on Innish Errich. He took her to the isle, and immediately left her there.

'In the woods around Innish Errich, he met Arindal carrying a royal stag over his shoulder, and accompanied by his five hounds. Arindal overcame Erreth, and bound him by thongs, cut from the deer hide, to a stout oak tree. Arindal set out to rescue his sister.

'Meanwhile, Armar, having left the island, had arrived at Daura's father's house. Realising what had happened, he called to some of his followers to bring more canoes, while he and the old chief went off at once.

'Nearing Innish Errich they saw Arindal in his canoe. Armar, thinking it was Erreth, shot him with an arrow. Arindal dropped dead, and the canoe drifted away. Armar, a very strong swimmer, hearing his betrothed weakly calling, jumped into the freezing water and, in the strong current, took cramp and drowned. Daura collapsed and died.

'Two weeks later a hunter, coming through the woods, saw a figure, white with snow, a stiff cold corpse, leaning against a stout oak tree. The flesh from its naked limbs was eaten away by foxes, and its eyes eaten out by a black raven, which the hunter scared away.'

It is a wild story and bleak. But now listen to the flora of the district, and know that it is not bleak, or necessarily poor. at all.

For here may be found columbine, Iceland poppy, lady's smock, Dutchman's breeches, common violet, ragged robin, shining cranesbill, wood sorrel, birdsfoot trefoil, hairy tare, tufted vetch, meadow sweet, cloudberry, raspberry, marsh

cinquefoil, aplune cinquefoil, strawberry, water avens, alpine
lady's mantle, wild rose, bird cherry, mountain ash, rose-root,
English stonecrop, pennywort, meadow saxifrage, yellow
mountain saxifrage, mossy saxifrage, purpose and golden
saxifrage, grass of Parnassus, sundew, purple loosestrife,
water purslane, willow-herb, enchanter's nightshade, wild
angelica, hogweed, giant hogweed, snake-root, hop, wych elm,
common rhododendron, yellow azalea, cowberry, blueberry,
cranberry, wintergreen, crowberry, ash, forget-me-not, ivy-
leaved toadflax, figwort, monkey flower, foxglove, brooklime,
water speedwell, red rattle, lousewort, yellow rattle, wood
cow-wheat, eyebright, pale butterwort, water mint, common
thyme, self-heal, red dead-nettle, white dead-nettle, ground
ivy, skull-cap, wood sage, bugle, bluebell, canadian elder,
guelder rose, honeysuckle, valerian, ragwort, groundsel, colts-
foot, dwarf cudweed, cat's-foot, daisy, marguerite, melan-
choly thistle, spear thistle, catsear, marsh dandelion, bog
asphodel, Solomon's seal, star of Bethlehem, wild garlic, toad
rush, hairy woodrush, daffodil, yellow flag, lesser twayblade,
bog orchid, frog orchid, fragrant orchid, small white orchid,
great butterfly orchid, lesser butterfly orchid, easy purple
orchid, moorland orchid, spotted orchid, northern fen orchid,
great reedmace, common cotton grass, deer grass, bulrush,
sedge, reed, sheep's fescue, meadow grass, crested dog's-tail,
false brome, couch-grass, Yorkshire fog, soft grass, wavy hair
grass, tufted hair grass, brown bent grass, common bent grass,
cat's-tail, timothy grass and meadow foxtail . . .

From Oban, where we spent little time, we took the car-ferry
to Mull, and drove down the long island to Fionnphort to be
ready to cross over the short sound to Iona by the first boat
of the morning. We found rooms in a modern boarding house.
It had no licence for the consumption of alcohol.

'It is very clean,' I said to Paul.

We drove to a licensed grocery and bought ourselves a
bottle of whisky each. Back in the boarding house a tinkling
miniature gong summoned us to eat. As we entered the room
we were formally introduced to the other guests – two couples,
who showed as little interest in us as we did friendliness to-

wards them. There were four very small flying ducks on the wall and a very large NO SMOKING sign. The bread was white and cut in triangles. We had some watery soup. Halfway through the meal a man got up and left:

'He's got something wrong with his stomach,' his wife whispered loudly to the rest of us.

It was difficult to converse; I made no effort.

'What,' asked the woman of the other pair, 'are we going to do next?'

'Roast-beef,' said the landlord.

It came out of a plastic packet; the mashed potato came out of ice-cream scoops; the cauliflower was white and soggy like the underneath of a bar of toilet soap. We had tinned pears afterwards, and we reflected that the entire meal, except for the additions of water, had been packeted, frozen, dried or tinned. Paul and I declined coffee, and drove down the road to find a bar.

We found a bar: cold, dark, dank and stinking of strong male urine. This was in an extension to a hotel; and the hotel had once been a decent building. We drove on another fifteen miles or more, from Bunessan to Pennyghael, where we had some glasses of Glen Morangie and cheered ourselves up.

Our drive back to Fionnphort was worth the excursion. The road runs alongside a sea-loch which was smooth and pale, partly pink and partly very light turquoise, a combination of colours I do not remember seeing on water. The fiery blood-orange sun suddenly, as we drove fast towards the southern tip of Mull, leaped out from behind the black cliff side of the banded mountain across the sea-loch.

Mull I already thought to be a most beautiful island.

All through the brilliant June day the sea and sky and rocks and wild yellow irises combined to dazzle and delight; and the only uglinesses were the broken-down crofts, the brutal squat churches, old petrol pumps – has anything beautiful come from oil, is not oil the great pollute? – the modern extensions attached optimistically to hotels which were the right size before; and the sense of dispiritment of the people.

This sense may not be accurate. It is difficult to tell. They only talk of the beauty of the weather, not of the beauty of the place. Trippers by coachload cross over Mull to take the ferry-boat to Iona. They pay their fares, inclusive round-trip fares, to MacBrayne's of Glasgow, in Oban. They are shipped to Mull and coach-loaded to Fionnphort and ferried to Iona and then fetched back again, all by MacBrayne's of Glasgow; and I doubt if any of them spends a penny in Mull or on Iona, unless it be two or three for a picture postcard or a cup of tea. They cannot be good for Mull.

But what is good for Mull? What is good for these lovely islands in the north-west seas? Even on the brightest, bluest, sunniest of days, such as we have had, with no cloud in the sky from dawn to dusk, there is a sense in which these islands are wringing wet with a kind of sadness. They are places which people who loved them have left, and may in their dreams 'behold the Hebrides' but not in their waking.

I had longed to visit these islands; and was filled with an inner joy at being amongst them. Why I should have had this longing I do not yet know; and why, having come there, I was not disappointed (as I usually am with places and people) despite the food and the bars stinking of urine and the inbred silences of the bred-out people, I have not yet ascertained. Yet I cannot, although I would were Paul to have sought to make his strongest attack, deny that there was about this trip, on my part, something of a pilgrimage.

It was so before I went, it remained so during the trip, and it is still so, in retrospect; more than a revisitation to Loch Awe; and more, too, than a treading of the route of the Harry Lauder song 'The Tangle o' the Isles' which my grandfather used to play to us, before the war, on his mechanical gramophone whose scratched and cracked notes I can hear now: 'Oh the far Cuillins are calling me away as step I with m' crummock to the road . . . Sure by Tummel and Loch Rannoch and Lochaber I will go . . .' I was brought up to be proud of the Scottish part of my blood, which came to me on my mother's side. It was really border blood: she was a Wood and her mother a Tennant, and her relations were Carmichaels and Mackenzies and Cowes and Christies and they tended to spread north from Berwick no farther than Leith and southwards to Newcastle. Thus there was no familial pull to the islands; and since my father's family were from the North Riding, I was very clearly north-east bred as well as Geordie born.

The advantage of being at Fionnphort so as to catch the first ferryboat to Iona easily outweighed the spotless and packeted discomforts of our lodging house. The first ferryboat crosses over the narrow sound before the first MacBrayne's coachloads arrive. We were able to walk through the narrow sea-facing village of Baile Mor, then on to the grass up to the Abbey without being part of a babble of trippers, whose noisy observations on this and that would undoubtedly have irritated both Paul and me, quite destroying the chance of any spiritual or historic apprehensions. Some fools will account this a snobbish observation but it is not so. For me to reflect or brood, and I presume for Paul to pray or worship (but he must speak for himself in all matters of superstition), is as impossible in the midst of a MacBrayne's coachload as it would be for a musician to enjoy a piece of chamber music in the middle of a Hampden Park crowd or for a man who loved his fellows a pint of beer in an alcoholics' ward, or come to that, for a healthy man who was not an idiot to perform the sexual act in public.

Also, I do not myself trust corporate or social emotions, being ashamed and fearful of the tears that fill my eyes at the sound and sight of men marching to martial music or of promenade concerts – I once went to the 'Last Night of the Proms', and was dismayed at the public rite. Such things as these are near enough to Nazi rallies or Roman triumphs. I am very conscious that privacy, which is not to say solitude, is in general the seemliest condition, and the one which is essential for reflection. This is a small part of my objection to churches, for their priests and spokesmen extol corporate acts and public worship and prayer, and indeed men who believe in Churches believe, supernaturally as well as superstitiously, which is to say ideally and ideologically, in the corporate existence and lively being of the Churches they have imagined and have forthwith joined.

On the way to the Abbey, Paul remarked upon a rather

scruffy plot of cabbages or some such, saying: 'I must say the monks, or whatever they have here, are rotten farmers.'

I said, 'I thought monks were supposed to be good farmers, having little else to do.'

'Monks pray,' said Paul, in terms of a rebuke.

'I once got drunk on green Chartreuse,' I replied.

We walked around the Abbey, restored decently except for the introduction of some hideous modern stained glass and a statue in the cloisters which has this inscription on it: *I, Jacob Lipchitz, a Jew faithful to the faith of my fathers, have made this Virgin for a good understanding among all the people of the earth. That the spirit may reign.*

A blind lamb emerges from a coiling mass of meaningless matter, so we are informed by an interpretatory leaflet, which goes with the large piece of cast bronze, and from which I quote: 'Lambs belong to the material world. Their only reason for existence is to live in order to suffer and be killed so as to feed the bodies of men. A lamb is thus a "symbol" of this blind suffering throughout the world of nature . . . Above the lamb, there is the Virgin. She is the symbol of the highest expression of Matter, a human person. Yet only her hands are fully human. She is capable of making an offering. But she is as yet blind (no eyes) to the full meaning of Life.'

Above the Virgin, indeed ostensibly supporting the Virgin, in a heart-shaped sack hung from its beak, is a Dove. '*Comes The Descent of The Spirit.* The Dove has large eyes meaning that while nature is blind (the lamb) and even humanity is blind (the virgin), it is by the Holy Spirit that we see. The Dove powerfully pours forth the Holy Spirit to envelop and to give Light and Hope to the Whole world of matter. The Spirit of the Dove pours itself down and returns upwards to invade and purify the Representative of humanity (and of all nature) in her body, through her heart, up through her mind to return whence it came.'

I walked around this sculpture several times, savouring each time the inscription, 'I, Jacob Lipchitz, a Jew faithful to the faith of my fathers, have made this Virgin . . .' and noting, in passing, that in *Behold Iona*, which could be regarded as the official guide, it is written: '*The statue of The Descent of the Spirit* in the middle of the Cloister by Jacques Lipchitz was erected in 1959.' To prefer Jacques to Jacob is to express a prejudice, as is the 'I, Jacob Lipchitz, a Jew.' In Scotland anti-semitism lies very near the surface, and it is often religious in its expression.

We mooched around the Abbey, both of us reluctant to leave it. Paul snatched a prayer while I took some photographs. I felt a bit scruffy doing so, particularly when I inadvertently caught sight of Paul praying; and remembered (and subsequently told Paul about) the occasion of the lying-in-state of Pope John, when Fergus Cashin (show-business chronicler of Liz Taylor and Richard Burton, Shepperton drinking-man, stage Irishman, loud and intelligent, and secretly unfully-lapsed Roman Catholic) and I had joined the queue of nuns and the like filing past the body. I had gone ahead of Fergus and had passed the cadaver and was half-way down the room when I looked back to see how Fergus was getting on. He had bent to kiss the Pope's slippered foot and was looking across at me to ensure that I did not catch him at it. Our eyes, that instant, locked on each other, and I could not but grin, nor could Fergus do anything but look shame-faced.

We made a half-hearted attempt to climb the small hill above the Abbey, and to inspect some of the mounds that lay beside it; and we had coffee in the Abbey café. I think I was more affected than Paul, and said to him, after I had acquired a great deal of the literature published and sold by the Community, that I fancied the idea of doing, or making (or whatever it is) a retreat.

'Retreats,' he declared authoritatively, 'are no good after the third day. I quite agree with what Jill Tweedie has been writing on the subject in *The Guardian*. They become exceedingly tiresome. I know, dear boy, for I have been on many.'

'I am not convinced,' I replied, 'for a retreat may be tiresome for people like you and Jill Tweedie but not necessarily for me.'

'People like you think they would like to go on a retreat. But you would hate it.'

'Why can't I bugger off? I would like to run away, to go where there is no telephone, no messages, no one to talk to.'

'You would be bored stiff.'

'I do not see why it is only Christians who are allowed to run away. When other people run away, they are called escapists, and the only place they can run to is the nearest lunatic asylum if they are poor, or the best nursing home or health farm if they are rich. Obloquy is involved; but it is praiseworthy to go on a retreat.'

'My dear George, you ought to join the Church.'

'Bugger off.'

'And another thing. You would be surrounded by all these women. It is dreary women, dear boy, who make up the great majority of people on retreats. They have lost their husbands and lovers, a kidney or a breast, they have not paid their rates, they have had their telephones cut off: they all have, they think, been grievously and unfairly injured by the world; and so they go on these retreats, not like you to get away from people, but because they are lonely. Look at them all, look around.'

He had a point: many might as well have been playing bridge or whist. A woman at that moment started a long tale of her injustice at the hands of the General Post Office.

By now, too, the coachloads were beginning to arrive; and we shook ourselves loose from the place, I armed with guide books and pamphlets and a long-playing gramophone record, one side, called *Coming,* being the morning service from Iona led by the Very Rev. the Lord Macleod of Fuinary, Founder of the Community, and the other, called *Going,* comprising a piece declaring 'Your holy hearsay is not evidence,' a song 'Hey, look outside' composed by Iona Youth Campers in 1969, and 'The poor are served' from a book of African hymns.

There was much of such goodwill to be seen in the Abbey. The young men cleaning the place were being good and smiling on their neighbours; and the coffee, although watery, was pleasantly served. I saw and see no reason why from childish simplicity as this I should exclude myself (and I would not be excluded, for they tolerate visitors of all faiths and of none); solely because I believe the beliefs of Iona to be false.

Their holy hearsay is not evidence, but is legend.

Walking back to the ferry, Paul and I spent a little time in the beautiful cultivated gardens in the ruins of the old Benedictine nunnery. Opposite is a shop tended by women, a Highland Crafts shop. Those inclined could buy there a Thistle Peat Reeker – 'Make your own instant peat aroma' – the contents of the tartan-covered box being: (1) Peat Vaporiser, (2) Vaporising jar, (3) Fuel, (4) Peat Powder concentrate, labelled Peat Reek.

Another kit offered: 'A part of Scotland for your home. Plant your own genuine Highland Tree. Contents: Pot made from Parent Tree. Bag of Sterilised soil. Packet of seed. Dibber. Instruction sheet.'

This part of the trip to the Western Isles is for many, and was for me, the nearest to a pilgrimage they perhaps, and certainly I, shall ever know: neither Canterbury nor Rome (nor Jerusalem nor Mecca) would end a pilgrimage. Iona is now, properly enough, within the Church of Scotland, having come into that society of presbyters curiously and circuitously: for the Columban monks that had followed in rough and ready measure the Columban rule for five centuries were ousted when the Anglo-Norman princess, Margaret, became Queen of Scotland and brought Scotland within the Roman Community and the Anglo-Norman influence. The Benedictines founded the Abbey, and this Abbey, which from time to time was also the Cathedral of the see of the Isles, was dismantled at the Reformation.

Iona passed into the secular ownership of the Macleans and then of the Campbells until in 1899 the eighth Duke of Argyll made over to the Church of Scotland the ruins of the Benedictine Abbey which, now restored, houses the presbyterian Iona Community. Iona's name has been with most Presbyterians since early childhood: and Iona, soft and sweet sounding, ancient and traditional, mitigated the fierceness of the dreadful protesting Scotsman, Knox, and the equally dreadful John Calvin of Geneva. Presbyterians and the Anglo-Scotch can have no early heroes. Anyhow, coming to Iona, I was glad that I had done so; and I recommend it to anyone else in similar plight, who feels inclined to make a pseudo-pilgrimage, or to make a retreat, or to make an escape.

The place is filled with good intentions; and I think it must have been so filled throughout its inhabited life, for there can never have been riches to be made here or armed power to be based here. Nowadays, what with its pacifism and its socialism and its preoccupation with race relations and its optimism about the work of the church in industrial parishes and the rest of the impedimenta of the well-meaning do-gooding sub-intelligentsia of the Christian church, Iona is no longer a place of intellectual, theological or moral distinction and exertion. But, then, why should it be? And, is anywhere? I do not think this matters. They take themselves seriously at Iona, they care for the buildings, they do little if any damage, so far as I can see: and this is more than can be said for most communities. They are civil, a little silly, spiritual vegetarians. If they'd have me, I'd retreat there.

There is something distasteful in the way in which I twist and twirl so that this joint trip undertaken following Paul's idea, becomes in my chronicling, an egocentric quest: but we cannot always choose where we purge ourselves. Also, I enjoyed myself continually in the far west and north.

Later in the year I drove with my wife and three of my sons and the eldest one's girl-friend to Cavtat, below Dubrovnik: Ragusa Vecchio it once was called. I had a discussion with my wife, who, seeing the many Adriatic islands off the Illyrian coast, declared: 'I think I hate islands. I do not see what men see in islands.'

I fear that what men see in islands is their insularity, their aloneness and their oneness. I've said it before, and so have others: John Donne was wrong, and all men are islands, whether they like it or not. No woman, or certainly no mother, is an island in the way that men are, and some strive to seem.

Later again, we came to Venice, island city, once immensely rich and very powerful. 'We might as well see Venice on the way back, then,' we said; and so we did. Amazingly beautiful of course, and amazingly run down, seedy. Beautiful, historical, and I gazed at it: but it induced in me no resonance. It is not northern and western, but southern, and eastern. There is no longing of my eyes for it whatsoever yet I know that it is a great place and the shell of a great civilisation. Nothing like this can be said of Mull, or of the entire Hebrides. Do, however, expatriate Venetians say that in their dreams they behold Venezia, as, in more dreams than were ever dreamed by those of Hebridean stock, did (and still do) men behold the Hebrides?

I, Y, pronounced Ee: the isle called by the shortest name for island, I. Iula, mis-transcribed into Iona, lovely mellifluous accident. Were it not that, had he not come and settled and sent his acolytes about proselytising, then someone else and possibly worse might have done so, I incline to the view that it is a great pity that Columba came to Iona and did not stay in Ireland, where he was bred. Had Christianity stopped there, in Ireland, and not infected North Britain, we would have lost the Lindisfarne Gospels and the Northumbrian renaissance, but there would have been the possibility of Britain, the Anglo-Saxons and the Germans and the Norsemen remaining decently and frivolously pagan, civilised by Rome, but not debauched by Christianity (which would have remained confined where it first festered, in the tideless unswept Mediterranean soup basin and among the feckless Romance countries, and up to but no farther than susceptible Hibernia).

Now, the Scots: consider them, or, consider us (depending whether you place an Anglo-Scot north or south of the border). They/we constitute a kind of ready-made and off-the-peg opera-bouffe in them/ourselves, whenever they/we dress them/ourselves in tartan kilts and begin blowing doleful pipes. And the Scots, were there more of them/us, might conceivably have gone the way of the Germans, for undeniably there is a streak of idealism flawing their/our entire system, a great fault the size of the Great Glen, a disastrous metaphysical propensity which not even the superb and splendid mind (quite unappreciated in Germany and France, where some think that Kant demolished him) of David Hume could put right. It is due to the paucity of Scotsmen or to their/our predilection to migrate from native hearths, or to the infirmity of their/our romantic, or idealist, resolution, or to the reasonableness of their/our best men, that Scotland has become since the end of the eighteenth century not some new thrusting Prussia to the fat, stuffed and slothful Rhineland and Austria of England, but a northerly appendage of northern England, following at a remove or two the London fashion, and a place sufficiently different from England to inform Englishmen in what their Englishness consists and sufficiently inferior to England to justify the English in our/their self-conceit. Scotland has fulfilled a great historic role in this regard, and also another role as the breeding ground of the policemen of the English Empire, now alas, defunct. They/we, the English, after all, knew perfectly well that they/we would never have gone out them/ourselves to manage all those tea and rubber plantations, or the great cotton fields, or to dig the mines for gold and diamonds, or to slaughter off Indians and Aborigines, or to convert the Kaffirs.

With passing mental indiscretions and fantasies such as these it pleased me to while away some of the holiday-brooding time upon rocks and at bars on Mull, and on grassy slopes above the Abbey at Iona.

111

Crossing back to Fionnphort in the ferry-boat I scribbled on
a brown paper bag which contained the gramophone record
produced by the Iona Community called *Coming and Going*
the following brief notes, which were intended at the moment
of scribbling to be notes for a poem called something like
Lines Written on Looking Back at Iona, but which had better
be called 'Going' (if not, 'Gone):

> *This is the last*
> *self-inflicted chewing – not by a swift sharp bite –*
> *through the umbilical Christian cord*
> *and making myself an island. From an island*
> *to an island to my island and even there*
> *I will be islanded from my island;*
> *chewing, natural and reasonable,*
> *yet not capricious as this sea and*
> *sky of hurried waves*
> *and the timeless ancient*
> *rocks.*

Much later, recollecting the emotions in tranquillity maybe, I
worked this over and this is what emerged:

MULLING ON COMING AND GOING

Not by a mother's swift sharp instinctive bite, end
Of birth and the pre-natal rule, this chewing and
Working through the christian cord takes time, and may
> *tend*
To separate us from beasts awhile. I'll pretend
We've will, thought thoughts, been more than chance
> *events' timed ground,*
And rejoice, in pretence, 'I, deciding, am not sad'.

From an island to an island to my island.
And here I have been islanded again. No land
Is firm, and firmament, enough. Chewing, reasoned,
I've thought things out, not capricious as this sea and
Sky of hurried waves is, shifting sea-weed and sand.

I've thought things out, and thought denies itself. Coming
To this beauteous spot was chance. I'm chemical. Going

112

Is the exodus of driven men. Dreams, homing
Longings: chemical; words: electric. Our sowing
Primes the fields, scatters good seed: energy proving
Nature's lush; seeding of fish and flowers: loving.

My love asserts to me I am no island. 'Sing'
She tells me as she sings herself. 'Be happy, bring
Joy!' Why not? Laughings of lovers, children, friends,
   string
All of us together. Happiness is good? Fling
Thoughts to the wind? The boat puts out. We're going.

Dr Johnson's mullings on Iona are much different, his
assertions and conclusions quite the opposite. The place drew
a famous resonance from him, and although the emerging
sound was quite contrary, a similar sort of bell was rung. But
what a different boom!

'We were now treading that illustrious Island, which was
once the luminary of the Caledonian regions, whence savage
clans and roving barbarians derived the benefits of knowledge
and the blessings of religion. To abstract the mind from all
local emotions would be impossible, if it were endeavoured,
and would be foolish, if it were possible. Whatever withdraws
us from the power of our senses, whatever makes the past, the
distant or the future, predominate over the present, advances
us in the dignity of thinking beings. Far from me, and from
my friends, be such frigid philosophy as may conduct us
indifferent and unmoved over any ground which has been
dignified by wisdom, bravery, or virtue. That man is little to
be envied, whose patriotism would not gain force upon the
plain of *Marathon*, or whose piety would not grow warmer
among the ruins of *Iona!*'

CHAPTER THREE

*Tales from Tobermory – the grand ruined Poor House
– Paul and I go to Church – 'Have you the Gaelic?' –
to Lochaline and Fort William – 'essentially a sporting hotel' – a dying derelict in a bar – and a Rhodesian
parallel – by the Pass of the Cattle to Applecross -
a strange religious place.*

We drove back again up the long island – and for any who
might think that driving up and down and across the length
and breadth of Mull is boring, it is not; it might become so
living there. But visiting is very pleasant, and the variety of
light more than compensates for the repetitiveness of driving
up and down, or across, its long and scanty roads.

The idea we had was to spend a weekend at Tobermory. On
the way we stopped off at a hotel chiefly made of logs:
Norwegian, modern, although the double-glazing had steamed
up. We arrived late for lunch – gone two, that is to say – but
they were very nice about it and we had a couple of drinks
while they made us soup and a cold plate.

The somewhat fey (in appearance and manner) but, we
jointly agreed, shrewd and clever lady who appeared to run the
hotel kept answering the telephone by the reception desk.

'Oh dear,' she said, after one complicated conversation,
'I've got the planes mixed up. I forgot to tell Prestwick about
it when they called.'

We were disconcerted at the implications of this remark.
She appeared to be Air Traffic Controller (part-time and very
casual) for Mull; and she did Air Traffic Control business on the
same telephone as the rest of her affairs.

A quarter of an hour later there was general excitement, and
people started running outside. We followed, towards the sea,
and saw an old red fire-engine being prepared. There was a
landing strip by the sea-shore which had been levelled as an

exercise by a team of army engineers. The small aircraft duly, safely arrived; and everyone, well pleased at the routine incident, returned inside. In Mull anything which actually happens tends to be an event; and everybody who happens to be at hand is entitled to participate in such events, if only by looking on.

At Tobermory, Paul was determined to go to church. I acquiesced. At the top of Tobermory there stood the newish Church of Scotland, removed from its earlier building down on the quay, presumably so that its more comfortably-off members, who lived highish up, would have less difficulty in getting to and from it. There was a Free Presbyterian church down at the bottom; and the arrangement was that Paul would go to it, so that he as a Catholic would discover what the other end of the spectrum looked like, while I, the lapsed Presbyterian, would refresh my memories of John Knox. In the event, Paul's nerve failed him, he fearing to go into a Wee Free place without benefit of guide or bodyguard; so we went together to the (Presbyterian) parish church.

Paul sang the hymns with lusty voice. He also admired the sermon, delivered with much competence and containing, so far as Paul or I could discern, nothing either heretical or offensive in the eyes of Rome. There was also (that which I had forgotten from my childhood) a children's hymn and a children's sermon, after which the children were led out into the church hall for their Sunday School. Before the departure the children had sung that sloppy hymn

> *O what can little hands do*
> *To please the King of Heaven?*
> *The little hands some work may try*
> *To help the poor in misery:*
> *Such grace to mine be given . . .*

This hymn is contained in the official presbyterian Church Hymnary in a sub-section called 'For Little Children'; I was instructed by one of my few intelligent school masters, Maurice Robinson, that the word 'little' had become so sentimentalised that it was best never used. When it came to the last hymn,

sung with great vigour, I barely prevented myself from
blubbing.

> *Glory be to God the Father,*
> *Glory be to God the Son,*
> *Glory be to God the Spirit, –*
> *Great Jehovah, Three in One!*
> *Glory, glory,*
> *Glory, glory,*
> *While eternal ages run!*

Oh, yes, we all remembered that one all right; and the sing-
ing of it; and my father's mumble alongside me; and my
mother's full-bodied soprano leading the choir.

> *'Glory, blessing, praise eternal!'*
> *Thus the choir of angels sing;*
> *'Honour, riches, power, dominion!'*
> *Thus, its praise creation brings;*
> *Glory, glory,*
> *Glory, glory,*
> *Glory to the King of Kings!*

When we stumbled out in the bright sunshine, we avoided
conversation with the Minister who, in his prayers, had seemed
to go out of his way to welcome and to pray for the visitors
from England in the congregation's midst.

Paul said: 'Have you noticed that almost all the congregation were women?'

As far as my experience goes, all congregations are mostly
women. But we were in little mood to discuss the experience;
and glad for some gin, back in the bar.

In the afternoon we drove across Mull, passing – for the
second time – what seemed to me to be a derelict mansion
suitable for restoration, to Calgary Bay, where, sheltered
from the wind, and warmed by the sun, we lay upon the sand;
then climbed a little up the turf and walked about, looking
across to the Treshnish Islands and Coll, to Ardnamurchan and
Skye.

These Inner Hebridean prospects in Moray McLaren's
indispensable *Shell Guide to Scotland*, are called 'in general, the

most spectacular in Scotland.' McLaren (any relation of the great gamesman McLaren, discoverer of Croquet's 'McLaren Peel', friend of Stephen Potter and the Simpson, whose son, Robin, now a lawyer, was among those who first introduced a new, sceptical Cambridge generation to the filthier devices of that peerlessly bloody-minded game? It might well be one and the same McLaren; in which case, should he ever read this, I offer praise and thanks twice over) also informs us that Calgary, Alberta, was not named after emigration from Mull but because a Colonel Macleod from Skye, Commissioner of the North-West Mounted Police, was related by marriage to one J. Munor Mackenzie who owned the Mull estate and because of which named the Canadian town after the Mull spot.

Calgary, nevertheless, has a deserted air; its beauty is lonely; and I would have preferred to believe that Mull islanders, choosing emigration to the Poor House, had carried images of lovely Calgary to the wheatlands of Canada where, far from sight of water and of the other Hebrides, they named their settling down place after the image they preserved, and sang their celebrated Canadian Boat Song: 'Yet still the blood is strong, the heart is Highland, And we in dreams behold the Hebrides!'

Alexander MacLean came to see us on the Sunday evening. He had just started a monthly newspaper, *The Western Isles*, and the day before, when wandering along the Tobermory harbourside, I had been taken by an announcement which said something like 'Alexander MacLean. Journalist. Knock and enter'. So I knocked, he said 'come in' and I introduced myself. He said he would join us the following evening and so he did, in kilt and the rest of the outfit, with a Lancashire accent.

He was born in Argyll, however, he swiftly said, and told how his father had run a department store and how he, too, had been employed there.

'I received my wages at the end of each month and my commission on sales each fortnight. At the end of one month, with all my pay and also my fortnight's commission I went off for a weekend and returned broke. I saw my father coming down the very grand and ornate staircase of the store. I went up to him and asked him to lend me ten shillings. He put his hand in his pocket for his wallet and was about to give me a ten shilling note when he said "But you had your month's wages and your fortnight's commission on Friday." "Yes," I replied. "Then you'll get no money from me." "But I'll starve," I said. "Then go away and starve," my father said to me.'

MacLean was a slight man, with a tidyish moustache.

'I went into the Air Force. At the end I was a squadron leader. I thought after it all, I would go into the snack-bar business. I went to learn the trade. I opened the vegetable trimming department each morning, clearing the rats out. I went to Charles Forte to seek his advice and he said to me: "How much money have you got?" I told him what I could lay my hands on, say it was about nine thousand pounds, and he said: "It's not enough. You'll lose the lot." I started writing a bit, back in Lancashire. Then four years ago, I came here. I have always wanted to come here.' We were drinking malt whisky in the bar of the civilised Western Isles Hotel.

Down the bar was Lord Maclean, chief of his clan, Chief Scout, who has since become a courtier. He lived on the island down at Duart Castle, and is Lord Lieutenant of Argyll, and opens fêtes and the like. He wore a kilt, but it was subdued; and above it he wore an ordinary hacking jacket, not the fancy garb that Alexander MacLean, our companion, wore. His *Who's Who* entry makes no mention of the 27th Chief of the Clan Maclean receiving any education, which, since he obviously did, constitutes a curious and possibly unique omission. He had been performing some public duty but instead of staying he settled down for gins and tonics: he did not necessarily do what was expected of him. To be Chief Scout in the sixties was to be brave.

Our MacLean, Alexander, small journalist, bravely had started up his own paper, wrote it, typed it out, stuck up the headlines with Letraset, and bundled it off to Oban to be reproduced. He was also one who had not done what was expected of him. He had left behind his prosperous prospects, had lost his money in the snack-bar business, had turned to scribbling in Lancashire – all this, routine enough – but had finally come to Mull, his Lancashire accent and attitudes alike overcome by his longing to return to live and work in his ancestral territory. Few Scots return, and few of those who do, return to work, and to start things. Most of them return to revisit, and look back, to tidy up things, to retire or to boast.

Alexander MacLean told us about the ruined place outside Tobermory which I had supposed to have been a grand mansion whose roof had been allowed to fall in and which had suffered the depredations of weather and looters.

'It was a Poor House,' he said. 'They were put in there from all over Mull and from all over the Hebrides. Whole families were put in there. The children of the families in the Poor House went to school in Tobermory, into the same classes as all the other children. But they were made to sit on separate forms. That was a terrible thing to do to them.'

I said to him: 'Why is the house not put right? The walls are good. The structure is handsome and finely proportioned. Much of the roof survives. It is a far better building than that Norwegian wooden hotel you've got, farther down the other road, beside the landing strip.'

He said: 'It belongs now to a Tobermory man and a solicitor in Oban, and they'd hoped to develop the site as a caravan camp. But the building, they say here, is now dangerous. There is a film company who want it. They want to burn it down for a film. That would be a very good thing for Mull. It would clear the mess up.'

'I think it will make a bigger mess. It could have been a fine building.

But its ruinous condition, and the knowledge of what it had been and had meant, made it ugly to Alexander MacLean, who was proud of Mull and who wished the best for the islanders, and who therefore wanted it burned down and the mess thereby removed.

He told us a complicated story about the Norwegian wooden hotel we'd had lunch at: how it arrived from Norway weeks late, how it had been photographed by a *Scottish Daily Express* man in an aircraft in its crates being carried down the Sound of Mull, how the bits and pieces had lain around after they were landed.

MacLean also told us of the Blackburn drivers' rally in Mull.

Blackburn had been part of his Lancashire connection; and MacLean wished to organise on a regular basis a Mull motor rally as a tourist attraction; something, presumably, along the lines of the Isle of Man's Tourist Trophy bicycle races. There is a rich and eccentric inventor called Dr Ochs who lives on Mull and commutes to London. Also on Mull there lives, when he is not in his Kent seat or his Antrim house, John Clotworthy Talbot Foster Whyte-Melville Skeffington, Baron of Loughneagh, Baron Oriel, Viscount Ferrard, Baron Oriel (UK), the thirteenth Viscount of Massereene and Ferrard, of Knock. Lord Massereene in his earlier days had been the driver of the leading British car in the Le Mans Grand Prix of 1937, was one of the original pioneers in the commercial development of Cape Canaveral, Florida, presented the operetta *Countess Maritza* at London's Palace Theatre, was Gold Staff Officer at the Coronation in 1953, writes articles on sporting and natural history, and recreations are all blood sports; farming; forestry; racing. In short, a splendid fellow (or so his *Who's Who* entry suggests).

'Dr Ochs drove me at 120 miles an hour down Glen More (which runs through the middle of the island) in order to prove that Lord Massereene was right in saying that if I had my bloody rally, you'd have cars going at a hundred miles an hour in the glens,' said Alexander MacLean.

He left us exhausted by his conversation and admiring his spirit, wishing that his enterprise be rewarded but hoping that this particular enterprise would be frustrated. He wished to get things done. This is not on the whole a happy wish to have in Scotland, and particularly in the highlands, and most particularly on the islands.

I drifted through the public bars of Tobermory, down on the quay. Here, as in Cornwall, the locals drank, many of them, as if it were their chief activity. In one bar by the harbour, to my right sat two old men, both, I'd say, having been fishermen. One was clearly more prosperous than the other, for he had motioned his friend over from the corner and was buying the drinks and was not so unkempt. From the way they talked, their relationship, notwithstanding, was of equals. They talked in English, until they were addressed in Gaelic, by a more educated-looking man on my other side, around the corner of the bar.

The educated-looking man might have been drunk – certainly, as they said, 'had drink taken' – and it was clear that he was asking the two old men what was the Gaelic for a particular word; and it was also clear that the two old men had no wish to become drawn into argument.

Eventually the educated-looking man said, in English, and indeed with an educated voice: 'You should say your mind, you should utter your opinion, not like the others do, who are silent, or who cannot say their minds. They do not know the poetry like we do.'

These remarks passed across my face and were impossible to ignore, not that they were intended to be ignored and not that I wished to ignore them.

The tidier of the old men then spoke to me: 'You do not have the Gaelic?'

'The Gaelic' is regarded as a possession, a thing of pleasure and of value to each man who possesses it.

'No,' he answered for me, 'you will not have the Gaelic.'

'No, I do not have it,' I said.

'He has been asking the Gaelic name for this month. He has been asking in his Gaelic what was this month's name,' the old man said of the educated man, who was now concentrating on his whisky.

The old man then recited to me the names of the twelve months in Gaelic and then said to me, with firm pride: 'I will give a pound to any man in this room who can say that.'

This surprised me; for they all appeared to know some Gaelic, and I would have thought that the names of the months were names learned very young and indelibly.

'This is beautiful weather,' the old man said; and I agreed, saying that I did not suppose that it would last.

'No,' the old man said, 'it is not like that in these parts. The more there is of the early sun, the longer will the summer be.'

The educated man now lifted his face from his glass and said to me: 'You do not mind us speaking the Gaelic?'

'No,' I said, 'why should I?'

'You know what we said.'

'I do not know the Gaelic. How could I know what you said?'

'But you are an intelligent man. I can see it in your face. You know what we said. You have heard it all before. You have heard this tongue before. Have you heard this tongue before?'

'Yes, I have heard the Gaelic before.'

'Tell me where,' the educated man asked, eagerly.

'I have heard it in the Outer Hebrides, on Lewis and Benbecula,' I replied.

'Aha! You know the Outer Islands; and Uist perhaps?'

'Not Uist,' I said.

'I am from Uist. Are you here for work or for pleasure?'

'I am with a friend. It is pleasure, but we are working.'

'And you will be going to Uist?' the educated man asked.

The old man laughed when he saw that I had shaken my head.

'He'll never go to the Outer Islands for pleasure,' the old man said to the educated man, who walked straight out of the bar. Straight, like an arrow: the way very drunk men walk when they grip themselves so that they do not stagger.

The old man turned back to me and said: 'It is an awful pity about that man. He is from Uist. He is a teacher. He is a very clever man, a very clever man, and he has the poetry. But he cannot leave it alone.' He tapped his glass of whisky. 'He cannot leave it alone.'

Woken at 5.45 on a Monday morning with coffee and toast, we left the Western Isles Hotel at 6.20 and drove through a crystal clear morning to Craignure and took the 7.30 ferry to Lochaline, whence to Fort William by 11 o'clock.

Fort William I remembered with affection, having seen some Highland Games there a few years back and watched the race to the top of Ben Nevis and down again. Here I first acquired, after some effort, the taste for straight malts, doing it the hard way, starting off with Laphroaig, which to the uninitiated can taste not unlike a disinfectant, but which leaves little hangover.

We used to have Laphroaig at home sometimes and would offer people 'the special malt, or the ordinary Scotch.' The greedy ones invariably asked for 'the special malt'. I'd give them the Laphroaig and they'd drink it with protestations of delight: yet most of them must have loathed it, for its taste is not to be acquired in an evening. It is a taste you must work to acquire, like most good tastes, and determination is required as well as application: unlike some of the easier and more fashionable Glens, or Ted Heath's Talisker. On this trip we were still sticking to Glen Morangie. It is a good malt. Although very easy the first time, it does not pall.

Fort William I did not recognise. Paul spluttered in righteous disgust at the cheap vulgarity of it all. There was neither trace of any elegance from an English settlement nor the orderly seemliness of a 'made' town such as Inveraray. It was all gift shops, miniature sporrans, etc., and tartans everywhere.

On the road coming, we passed, going the other way, a tourist coach which had stopped at a spot on the road for no apparent reason. Then out of the bushes, fully dressed and equipped for the part, kilted and tartaned and otherwise appropriately accoutred, sprang two bagpipe-playing 'Heelanders'. The trippers, clearly forewarned, did not bother to get out of their coach but took their pictures of the bagpipe players through the windows of the coach.

Paul tried on a tweed hat in a shop in Fort William and, peering into the Cellophane cover of a matching wool and skirt-length set (for the shop, although clearly both the most socially ambitious and also the best in the place, appeared to have no mirrors in which people could see themselves wearing their prospective purchases), thought it made him look ridiculous; and decided against the headpiece. It did indeed make him look ridiculous; and he also looked ridiculous looking at himself in the gift-wrapped Cellophane cover to discover whether he looked ridiculous.

Throughout the trip he has been wearing a blue corduroy peaked cap to his evident satisfaction and quite careless of its ridiculousness. Wherever we have gone, this hat or cap of his has caused us to look like a pair of freaks. Whenever I have mentioned this, Paul has suggested that I have not looked all that normal myself. We agreed on hating and despising Fort William. After Paul had got some money we left.

We had thought to have lunch at the Invergary Hotel, well-enough recommended by the *Good Food Guide*, although this is a volume which I distrust since it claims rather than conceals that it is written by its readers. (A phrase from a *New Statesman* review by Jean and Andrew Robertson, a kind of journalistic double-act in the consumer business line of trade, quoted self-indulgently on the back of the cover of the issue Paul had brought along, says: 'As far as we know, no other country in the world has a guide that is virtually written and bought by the same nucleus of people', which is to my mind a damning judgement, but clearly not to the minds of the Robertsons and the publishers of the *Good Food Guide*. But if you know that nucleus of people, and they may best be described as *Guardian* readers leavened – or lumpened – take your pick – by Bon Viveur fanciers from the *Daily Telegraph* – if you know that nucleus you can use the Guide, although naturally not as Jean and Andrew Robertson enthusiastically intended.)

Here is an example, from the *Good Food Guide*, 1971, page 325. '"The Invergary is essentially a sporting hotel," says the brochure, but this refers to the stalking and fishing within easy reach, rather than to spartan conditions indoors. There is a dog register in the hall, a log fire in the bar, and attractive furniture, and flowers everywhere. Captain Hunt expects you to enjoy his French wife's cooking (which benefits from both cultures), and many people do: "delicious thick soup, full of flavour – rather as if last night's stew had been put through the blender"; "the prawn bisque was exceptionally good"; "the roast chicken was a tender bird, served with a stuffing of toasted oatmeal and onion, and a delicate egg sauce"; "good English (*sic*) [their *sic*, not mine: GSG] cooked with herbs and *fantasy*" [my *sic*, not theirs this time, and my italics, too: GSG]; "the syrup tart was nicely gooey." But there are complaints of less successful soups, over-herbed casseroles, poor vegetables, dried parsley, a meagre cheeseboard, and coffee tasting of instant. The service is usually attentive – rather too

much so if you linger after dinner. There are almost sixty
wines . . . Draught Bass and lager.'

This is the sort of informative writing that informs you
about the writer rather than his subject, and does so acciden-
tally rather than on purpose. It is none the less valuable for
that.

Some days, or it may have been weeks, earlier I had said to
Paul that I preferred public bars to saloon bars, they being
more interesting. This is the sort of loose and sloppy and
indeed generally dishonest statement which I am over-fond of
making. Having decided that it was, in fact, too early for
lunch, we determined that we would, instead, have a drink at
this remarkable hotel.

Paul strode ahead. He usually did. His natural speed of
walking is much greater than mine and amounts, indeed, to
something like a fast march. He pushed his head in at the
front door of the Invergary Hotel, pulled it out again, and
then marched off past a petrol station (closed) and a closed
snack-bar advertising soups, towards the hotel's public bar.
Trailing beyond, I made no protest; but rather approved of
this.

The hotel entrance had not looked inviting. I had not cared
for the 'dog register in the hall'. Also, the matter of Scottish
licensing laws was a continual source of concern to us, and of
occasional distress. The hotel looked as if it did not want
people to drop in for a drink, particularly if they were un-
accompanied by dogs. There was no reason, of course, why it
should look otherwise; or be otherwise. There are in this world,
hotels for all men and most dogs.

The public bar was a detached rectangular block, cheaply
and crudely built, a cold and plain box with a floor of linoleum
tiles stuck on to the concrete. There were a couple of tables
with plastic tops and small stools beside them. The bar itself
was too high to lean upon with an elbow, comfortably; it had
a Formica top. The walls up to a height of about six feet were
painted a pale glossy blue, and above a glossy cream. In the
corner above the bar an ill-made shelf supported the television
set.

There was a familiarity about the scene, and about the

relationship of this public bar to the hotel of which it was a part and an outpost. The set-up reminded me of the African bars attached to, or outposts of, some of the bigger hotels in Rhodesia. The noisy laughter, violent language and coarse buffoonery of the workers drinking here in this Scottish bar were not dissimilar to the noises of the blacks and the poor whites in the Stanley bar of Meikles Hotel, Salisbury, Rhodesia.

Rhodesia is very largely a Scottish creation, made by the grandfathers and fathers of men like these in the public bar of the 'essentially sporting' Invergary Hotel. If the men in the public bar were to go to Rhodesia they would go to the Whites Only bars (for the whites in the mixed bars in Rhodesia were either poor whites, or journalists and the like), and they would suddenly have become the bosses. They would remember the lairds, and they would feel superior in Rhodesia just as they had felt inferior in Scotland; and they would imitate the attitudes of the lairds as they had seen those attitudes, and within the Rhodesian context: and, as they, or more often their grandfathers and their grandfathers' grandfathers, had been colonised in their own country by the lairds, so would they in their turn colonise the country of the 'Kaffirs' and make it as much their country as their old country had been made the country of the lairds. The Scottish lower classes who colonised Africa learned their colonial manners from the Anglo-Scottish younger sons of the upper classes who colonised Scotland.

In the public bar of the Invergary Hotel, where the *Good Food Guide* told us they did good English food and that the syrup tart was nicely gooey, a notice said: 'We regret drinks cannot be served on credit.'

In one corner a man sat crouched over the end of his drink and the end of a cigarette he had rolled a few minutes before. His body heaved and shook as he coughed again and again. His face was almost white, and the bout of coughing and retching made him shake and sweat. At almost the end of his coughing fit, phlegm and spit dribbled out of his mouth and broke into three separate trickles falling down his chin on to his pullover.

He was a young man, and evidently close to death. He could very well have been a consumptive African in a beer house in Rhodesia; but he was a broken-down piece of Scottish refuse in this public bar in Invergary, a finished man, oblivious of the savage laughter of the swearing, happy, healthy men from the hydro-electric scheme at the other end of the bar, exchanging maty words with the local policeman who had dropped by.

There are classes within classes down to the smallest of groups, and in the smallest of groups each man is an individual (or an island, until, like the finished, wrecked, dying man in the corner, he is drowned), however he may appear to the rest of the world outside as one of a group, or of a class, or a race, or whatever.

It is easier to get to Applecross by sea than by land; and in the worst of winter it is the only way to get there. I think I would really go there if I wished to retreat, instead, that is, of making a retreat. It would not be an embarrassment or an affectation to escape to this remote and pleasing spot, whose remaining years of habitation may well be numbered.

We drove to it over what McLaren calls Bealach nam Ba, and the Pass of the Cattle: the highest and fiercest pass we encountered. The village itself lines the sea wall and looks out to the islands of Raasay and Skye beyond; its immediate hinterland is arable and, coming down to it from the high pass of rock and peat and bog and lochan there is a sudden softening into fields and English clumps of deciduous trees. The name, too, sounds English. These superficialities apart, the place is of the greatest Scottish purity.

In the evening, before dinner, we walked, then returned and stood on the narrow road between the cottages and the shingle as the air cooled and the sky slowly dimmed. A woman who was another visitor, standing not far away and looking across the sea to Raasay and Skye, turned and glanced at me.

'A peculiar and a beautiful place,' I said.

'Yes,' she replied. 'Do you realise that they have three or four churches here although I do not suppose there are a hundred inhabitants.'

I assumed that she meant that there was an 'official' Church of Scotland, a Wee Free body, and possibly a handful of Romans.

'This is a very strange religious place,' she said. 'Here, they only take Communion once a year. If they think you're a backslider they won't let you take Communion at all. The Communion services of the three churches go on for a whole week, from Wednesday until Wednesday. They dress all in black. They take each family, one at a time, on its own and each family's Communion service lasts for four hours.'

'How do you know all this sort of thing?' I asked the English woman.

'My sister-in-law was and is an elder of one of the churches. My brother-in-law joined the Anglicans because he wasn't going to have my sister-in-law sit in judgement upon him. They have different degrees of strictness, you see, and if they think they may be refused Communion they change to the less strict Church. There are three parsons here and there is no pub.'

'What do the men do for a pint?'

'There do not seem to be many men around.'

'It is an odd place.'

'It is very odd. I like it. It interests me,' the woman said. 'I come here each year, for the quiet. I quite like reading. But, mind you, to sit and read poetry, or prose – oh no, not that, I couldn't bear it.'

Apart from walking around the long straggling line of cottages, and discovering three chapels at one end and another at the other, and noticing the people dressed in black and the scarcity of men, I did nothing to substantiate the English woman's hearsay. I liked the sound of it as it was, and it was what she believed to be the truth about Applecross, where, according to McLaren, 'In A.D. 671 Maelrabha, an Irish monk from Bangor, founded a monastery in this place. It existed for about 100 years before being destroyed by the Vikings. Remains of a pre-Reformation church can still be seen, and an old cemetery containing a Celtic cross 9 ft high.'

The preciseness of the dates which guide books chuck around always amuses. The knowledge of places and dates of the seventh century is scanty indeed – far less accurate than the

hearsay information supplied to me by the Englishwoman –
yet I would not take her statements for fact, but only for
evidence of what she believed to be true. 'An Irish monk from
Bangor' creates a quite false impression, for in no useful sense
was there Irishness in the seventh century and in no sense was
there Bangor, and only in our senses were there years called
671 and people named as monks.

The three churches at one end of Applecross are plain and
simple buildings, each with its own notice-board, each giving
the times of its service, each secure enough in its whitewash
and railings. There is nothing strident about them, nor any
sense in Applecross of the rigorous moral striations that their
Communion practices delineated, if the Englishwoman was
accurate in her remarks. The solitary church at the other,
northerly end of the village, beyond the entrance to the
principal local sporting estate, is beautifully restored within.
Barn-shaped, the walls bare part-dressed stone, there is a tall
free-standing pulpit with its built-in staircase standing erect
and tall in a corner and a communion table below the central
window; a very satisfactory building, alone in its cemetery with
a few trees around and grave-stones among the wind-bent tall
grass.

We stayed at the small hotel next to a licensed grocery; and
there was a beautiful, slightly accented young woman with
expensive clothes who had driven herself to Applecross in her
Mercedes sports car who scarcely ate any food at dinner,
toying nervously. In the morning, at breakfast, she re-
appeared and now had become far calmer. I do not know
from what she was escaping, but I am sure she had fled north
and that Applecross had comforted her.

Applecross is a detour: a cul-de-sac. We drove back and up to
the high plateau, Applecross and Skye behind us and the Pass
of the Cattle below us. We sketched on the top – and there was
the smattering of the first rain we had had. We dropped down
the Pass and drove on to Shieldaig and Torridon, viewing the
smooth white and shining quartzite Torridonian peaks of the
red sandstone mountains. This stone looks, and is said to be,
very old. The clouds and mountains obscured each other.

'This,' said Paul, 'is real Highland scenery, all rain and clouds and mountains and mist, and sodden sheep lying in the middle of the road.'

Our sketches on top of the Pass of the Cattle had been entirely different – and Paul's much more confident than mine. I'd also taken some photographs for subsequent fortification.

Paul said, looking at both sketches, 'It's very interesting how differently people see the same things.'

If the sketches we'd produced were anything to go by, it would indeed be interesting. However, men assume, without much evidence except for commonsense, and presumably without a possibility of proof, that we *see* more or less similarly but draw it differently. I do not think there is much more to it than that, although there is certainly something more: when men were frightened of the Lake District their drawings of what, to us, are gentle, easy and softening views become fierce and harsh and frightening. We both could see, driving over this pass to Applecross, how full of terror such a landscape could be.

Earlier, I had seen out of the edge of my eye a slab of rock

high above, on a steep mountain side. Above the rock, the wet
earth oozed a film of water which slid over the rock face permanently: there had been no rain for many days but still the
rock face shone with the sweat of the earth above it. Glancing
out of the edge of my eyes to the dark mountain hanging
above, the sun suddenly caught this rock-face which, as a
reflecting mirror, chanced to flash the sunlight back to me as
we passed. Any man travelling with difficulty on an old cattle
track, alone or merely with a hired servant, and possessed of
and by a mind filled with convictions of the supernatural,
might well have been instantly and permanently convinced
beyond all doubt that he had seen a mighty flashing shield, or a
flaming sword buried in the rock, or in like but different circumstances have seen and heard instead of a shaking tree, a
headless cow yelling, or a headless dog howling, or Christ
resurrected, or any blinding light.

Later, when we stopped for lunch, I began scribbling a note recording this unremarkable observation.

Paul said to me, 'What are you writing? A prose poem?'

'Something like it,' I replied. 'I am trying to put you off. You keep writing all your bloody little notes down while I am doing the driving, so now I'm writing something down.'

CHAPTER FOUR

*To Shieldaig and to Torridon – the different ways the
same scene is seen – a hot-water bottle at Inchnadamph
– and a bore – and eating there, especially pink trout
for breakfast – Cul Beag, Stac Polly and Cul Mor –
the sight of the northern sea – our lucky escape from the
Arts and Crafts Centre at Balnakiel – the Kyle of
Durness and the Cape Wrath hotel – the man from
Humberside and the Public Bar.*

We drove through the Glenshieldaig Forest to Shieldaig and
by Loch Torridon to Torridon and through Glen Torridon to
Kinlochewe. This is a road which some claim to be the finest
in all Scotland.

We had driven on from Kinlochewe northwestwards by the
southern bank of Loch Maree which some say to be the finest
of all Scottish lochs, and then to Gairloch and Poolewe,
skirting the sea of Loch Ewe and the sands of Gruinard Bay
and inland and southeastwards again by little Loch Broom to
the Braemore Forest and Gleann Mor and north-northwester-
wards by Strath More and the north-eastern side of Loch
Broom to Ullapool; and this, too, is a most fine route, which
lovers of Ross and Cromarty will rightly commend.

And Ullapool was good to look at: a place in its own right,
not existing to service passers-by, a fishing place, a broad
street, whitewashed stone.

But Ullapool became, in retrospect, something else: for it
was the beginning of a piece of road and land which were to us
swiftly to become, as they already were to those who had
known them long and had compared them with far more
knowledge than ever I could, the finest roads and lands, the
finest mountains. lochs and seas and skies that are to be found
anywhere within the British Isles; or, indeed, anywhere in the
world.

Although in size this piece of land is small and in scale also, yet the relation of land and air and water, overcast with the particular natural and human history of the extreme north-west mainland of Scotland, is such that not the Himalayas around Kathmandu, not the Alpine passes between Switzerland and France and Italy or between Austria and Yugoslavia, not the stupendous clefts of Africa or the Rockies, and not even the better known and loved Lake District of Cumberland nor the beaches of Northumberland, can any more compare.

Kashmir, I found beautiful, and very moving also: but, I have not found anywhere on earth more beautiful than the western lands to the north of Ullapool, nor do I expect to find anywhere more beautiful, or hope to.

In this verdict I am in good company: Compton Mackenzie writes: 'If I am asked what I consider to be the pick of Scotland's scenery, I shall reply (with apologies to Loch Sunart and Moidart, to Arisaig and Morar), "From Gairloch to the isolated bens of Sutherland". Of all the lochs in Scotland, Loch Maree is most often in my mind's eye; of all the bens, Suilven. And, if it should happen to have been wet in the month you chose, the water-lilies in the lochans will make you glad that you did take your holiday then.' I have placed my ideal slightly to the north, excluding Loch Maree so as to include

Ben Loyal, the Kyle of Durness and Cape Wrath. Paul and I were doubly fortunate; for it did not happen to be wet in the time we chose. The rain we encountered driving over the Pass of the Cattle and up to Inchnadamph petered out. Once more the clear and pale blue-green skies emptied themselves of all clouds. And still there were the water-lilies in the lochans.

It was from Inchnadamph; and then from Durness, that we moved around these mountains and lochs. Around Cul Beag, Stac Polly and Cul Mor and round by Lochinver and Suilven, by Loch Assynt, and round again; and then by the Kylesku Ferry to the north-west end of the land, and across the Parph to Cape Wrath. Throughout these days visual excitement was intense, and visual delight.

We spent a night at Inchnadamph, the longer to enjoy these splendid parts. The evenings now stretched towards midnight, the water was soft, the malt whisky smooth. To bed: and, delight! An unexpected hot-water bottle, unnecessary but welcome, a solace for my feet.

'We are now in the hot-water bottle zone,' Paul announced, and he was to be proved correct for the following night we were to have hot-water bottles again at Cape Wrath.

The hotel yo-yoed in my estimation. I took to it at first and therefore unpacked, suggesting to Paul that since we were in danger of running out of Scotland on the north-western route, we might well stay here for two or three nights, and he did not dissent. The people were old. The place was fishy and supposed to be – there were great fat trout stuffed in cases. There was a terrible bore in the bar, full of misinformation and prejudice.

'He doesn't know *anything at all* about what he's talking,' said Paul, in furious mood, after we had heard him on the subject of tied cottages and French Prime Ministers.

The bore contended, his voice full of shocked irony, 'It's wrong for farmers to tell their workers to leave their cottages when they're fired but of course it's all right, isn't it, old man, for the Prime Minister to leave Downing Street when he's given the boot. And I don't suppose he pays a fair rent either, when he loses his job, and he gets a pension – that'll make it now, er, four ex-Prime Ministers all getting pensions. And it's worse in France, it's a damn sight worse in France I tell you,

or used to be. I can't be certain of my facts now, but they
changed jobs every ten minutes and every job had a pension,
and those damned French politicians used to go around col-
lecting pensions. But tied cottages. If a farmer tells a worker to
leave, he's in trouble. The socialists have been seeing to that
for forty years.'

'He's a bloody *mine* of misinformation,' hissed Paul, near to
apoplexy.

'You must admit that his opinions are sound,' I said.

We didn't dispute that this man had been boring England
and Scotland for at least fifty years. His voice was also pene-
trating. We had difficulty placing him. He had a strong regional
accent, but great social assurance also. He was too wealthy to
be an academic and too stupid to be a businessman and not
smooth enough to be in law.

'I think he's a great doctor, a famous medical professor,
always bullying the nurses and shouting at the students at one
of the London hospitals, when he's not conning his Harley
Street patients with bullshit,' I suggested.

Paul at once agreed.

We were gonged, again, for dinner. None of this southern
affectation of going in to eat more or less when you felt like it.
We were put at a table with four others. I hated this. Paul was
far better at it, politer, less annoyed. Eventually, thanks to
the need to pass round the vegetables, we exchanged some
small talk and I found myself looking at a pleasant woman
with a startling resemblance to my aunts Winifred, Dolly and
Elsie. She came from Yorkshire, too, and could very easily
have been a niece of theirs and thus some kind of cousin of
mine. She had my father's and his sisters' light blue eyes and
fine hair and very thin skin, and her face was boned like theirs,
especially Aunt Elsie's, and even the way her face moved
when she smiled and her lips when she talked; and also, after
dinner, I looked at her walk and her legs and still she seemed
like them. I should have questioned her a little; she'd have
liked it, whatever the result. But I didn't.

The food to our surprise was good. 'These are the first
genuine potatoes we've had since leaving England,' I said to
Paul, and the rest of the tables simpered.

'The soup was good,' Paul announced. It had been made on the premises.

The meal was not perfect. They didn't bother with wine, not that that bothered me. Before the chance to ask for it arrived, the waitress came and said, 'Breakfast at nine?' looking at Paul and me.

'Is there any choice?' I asked.

'How do you mean?'

'Are you asking us or telling us?'

'What do you mean?'

'Is breakfast at nine?'

'Yes.'

'And not earlier?'

'Oh, you could have it earlier by special arrangement, I think.'

I was tired. She pressed on:

'Porridge or cereal?'

We all said porridge.

'Lovely,' she said, 'I don't have to think. It's six porridge. And bacon and eggs, all right?'

'All right,' we all grunted.

In the morning, before departure, while I was sitting in an upstairs lavatory, Paul badgered me to get *moving*.

The mail van brought the breakfast rolls, at about 9.15: they dine early and breakfast late in the Highlands.

'They keep to their winter hours all through the summer,' I complained to Paul.

We discussed this and kindred matters over breakfast with the regular couple at our table, the wife being the one who reminded me of my aunts, and we established general agreement as to the idleness and fecklessness of the average Highlanders and Islanders, as Englishmen abroad are wont to do.

We also generally complained of the frozen food, the processed bread from Glasgow and so forth. I remarked that although they had wanted to know each previous night whether we wanted porridge and some of us had said that we did, the porridge when it came was processed muck.

Portsonachan, to its great praise, remains the only place

where for breakfast, we were given proper chewy knobbly porridge which, cold, could be carved and which had been made overnight from oats. All the rest, wherever we went, was filthy processed pap, and we might as well have eaten Kelloggs corn flakes.

This morning, however, there was talk that some trout had been caught the night before. The bacon looked like Wall's, out of Cellophane wrappers. Of our table, when surprisingly we were offered it, only I chose trout. It was not at all like the trout of frozen packets I had become used to, but was dry, light and salmon-pink and fried in oatmeal; and my two trout caught from one of the lochs the night before, were truly delicious. I suppose that I chose them to be different. At all events, I was the gainer.

And now, at length and at last, we had arrived at the last stretch of this extreme north. Having crossed the Kylesku ferry, driving along the road to Durness, at a quarter to four in the morning, climbing above the Strath Dionard to our

south and east, we rounded a bend and saw, in the far distance, the sea: this not the sea to the west but the sea to the north. To our right was Beinn Spionnaidh and Cranstackie, and behind Fionne Bheinn and Arcuil, after which hills, in the English form, the Duchess of Westminster had named her horses Foinaven and Arkle.

Paul sketched. The wind blew the clouds along swiftly and the shadows rippled and dappled the great strath of land stretching towards the distantly visible northern sea. To the north-east the land fell downwards from Beinn Spionnaidh, and the foothills were bare rock with turf and heather between the outcrops. The land was bald right into the distance, down the land called Strath Dionard between Beinn Spionnaidh and Fionne Bhein stretched endlessly, bare and deserted into the unpeopled treeless wastes of the Reay Forest. Men had cut the forest down, to cultivate the land; then the men had brought in sheep and driven out the men; burning the bracken to produce spring food for the sheep had further ruined the land; and now there were barely any sheep either.

The land looked elemental and ancient; and indeed the rocks were ancient. But the bleakness was not as it would have been had men not lived and roamed and fought and grabbed and marauded and pillaged and been predators. The elemental look, the look of the wilderness, the ancient look: this look was recent and man-made. This treeless waste land, if left to its natural self, would have been forest, even with ash groves, as we had seen a tiny ash grove, miraculously preserved, in the Inverpolly Nature Reserve, hard by what now, at home, I see I fell in love with: the mountains called Cul Bearg, Cul Mor and Stac Polly.

Durness was calm, almost mild to look at: the landscape, that is. The houses remained bleak. And quite unexpected, the Kyle of Durness. With backs to the sea, we faced south, the sea from the north coast behind us, now straight to arctic waters.

The Kyle, a sandy-edged great estuary, flat with changing shallow banks, reminded me of Laugharne in a way – the sea proper not visible, and here even stranger than Laugharne, for the sea enters by a narrow neck by the ferry to the minibus that takes you to Cape Wrath itself, over the Parph. We delayed going across the Parph to Cape Wrath.

'Senses become blunted,' said Paul: 'we have seen too much scenery, and besides, I want a drink.'

So, come to that, did I. Paul went off to write some notes. I discovered there was a bar at the back of the hotel.

'You mean, I have to go outside the hotel in order to get to it?' I asked the maid.

'That's right, sir,' she said slowly, adding, as if to an idiot: 'You have to go outside, through the front door there, turn left, and continue walking around the hotel until you come to a door marked "bar" and that will be what you are looking for, sir.'

When, later, we made the expedition to Cape Wrath we saw, within half a mile of the Cape, standing alone a few hundred yards away a hundred feet or so below us, in the great un-inhabited wilderness called the Parph which sometimes the German navy shells, a large stag, easily the largest and most magnificent of the wild beasts of Britain.

Earlier, we had looked at, marvelled at, Smoo cave.

'Very fine cave,' said Paul, and so it is, great natural vaultings in the Gothic style, and deep slimy greens and pinks and mauves, all of that almost fluorescent and lifeless colour that seems to grow in dark wet places always out of the reach of sunlight. The rocks in the bucklings were finely cracked, as in slate, in bands of about one and a half inches wide or thereabouts.

We had also laughed aloud in self-congratulation when we saw the Balnakeil Arts and Crafts centre, for we had thought to stay here, not knowing it was a collection of old Air Force prefabricated barracks containing two or three men pursuing their crafts well, some others pursuing the art of painting very badly, and a coffee shop, closed. Paul, being left-wing, is more ideologically predisposed to Arts and Crafts than I, but even his theoretical loyalties to the principles behind the dump could not prevent him shuddering with revulsion at the practical end-product. We fled, joyful to have escaped; and I perhaps more joyful than Paul, for it had been my suggestion that, in the interests of local colour, etc., we might put up for a night in the place. We had indeed made inquiry on the telephone; but had been declined.

We were thus doubly fortunate to sit down to one of our best meals, dinner at the Cape Wrath hotel – leek soup, cold salmon trout salad, mutton, gooseberry tart. It was mainly local. We had a bottle of good claret. We more than once congratulated ourselves upon our escape from the Balnakeil Arts and Crafts Centre. All was contentment: except that Paul lit up in the No Smoking dining-room, prompting a man at a neighbouring table to say irritably and noisily to his companion: 'No, no, I *won't* smoke in here.'

In the bar, an old man said: 'There's money in the tourists. But there's nothing for the tourists. They're bringing the roads up here, all these new roads, and the traffic will follow the roads, the traffic always follows the roads, but where will they stay, what will they do?'

The bar began to fill, not with people from the hotel. A girl from the south of England said: 'My boy-friend, he's trying to get into the Craft Village. He's had an interview but he doesn't know yet. He's teaching now.'

The woman behind the bar said: 'If he wants to come up here, why doesn't he teach up here?'

The girl from the south, who wore pebble glasses and was frowsy and who kept her fat legs wide open, as if she was giving herself an airing, said: 'But he doesn't want to teach. He wants to do pottery, you know. He's hoping to get a grant.'

A man who sounded as if he was from the north of England said: 'No, I'm not from the north of England. We're from the Humber, and it's so black, the air there, we've got a biggish house, and all the brass door-knockers, you know, you polish them in the morning and they're black the following morning. Everybody feels like an old man there. I come up here, and after a couple of days I'm a young man again, the air!'

Another old and local man said: 'There's no pollution in the air here.'

'Right. It comes straight down from the North Pole,' said the Humberman, who now spoke directly at the old local man. 'We've come here from Africa on leave. Back to England, each two years. The first time we came back you could park the car. The second time, you could park the car but it was a little difficult. Now you can hardly park the car anywhere. Teeming with cars, the place is.'

I remembered at Inchnadamph the old bore saying: 'Things are not as they were, it's not like it used to be'; and here the same thing is said, but by a working man who'd made cash in Africa, who now said: 'A Ford car used to cost £90 and £10 to replace the engine.'

One of the old men said: 'Aye, a new Ford car, luggage grid, taxed, insured, ready to go on the road, a hundred quid dead. I said to the salesman, "Yes, I'll take it," and it's still going on the road, that car.'

Paul came in. He tends to enter rooms noisily, or, at least, since he doesn't necessarily make much noise, to give the impression of commotion and disturbance. His entry temporarily silenced the bar as they all observed him joining me.

'I've just telephoned Marigold,' he told the bar, 'and she tells me that a very mysterious and expensive fountain pen has just arrived at the house from the Western Isles Hotel, Tobermory.'

'That is my best fountain pen,' I told him. I was greatly relieved that the pen had turned up. The Western Isles Hotel, already high in our estimation, went up higher.

Soon the conversations resumed themselves. The Humberman on leave from Africa said: 'They're worth money now, you know, some of those old cars are, when you come to think. Did you see that thing on television the other week? The first driverless car. It looked like a Morris 16-20 to me and nobody in it. Man sitting in the back seat. A big box in it. Another ten years it will be a little box. This is the start of the new techniques. And there's pleasure to me in motoring. It's the same with ships, man – these big enormous tankers, in theory those great things can take themselves anywhere. They don't need sailors any more. That's the way it's going, man.'

Paul had surreptitiously removed from his pocket a tatty couple of paper napkins crumpled up. Inside them were a cigarette butt, some matches, accompanying ash and a filter. So ashamed was he of smoking in the dining-room that he had mopped up his nicotinal leavings in the paper and put them in his pocket. Now, in the public bar, round outside and at the back of the hotel, he fished out the detritus and put it all in an ash tray.

'Don't you put that down in your notes, George,' he says. So I do.

'Ha, ha,' he exclaims, 'I shall get you for that!'

The Humberman back on leave from Africa was getting talkative again: 'A friend of mine, the nicest man you'd ever hope to meet, there was this African came at him with a machete, hacked him. There was an African said to me, "Go on, hit us," as I'm taking him to the police station. I was all drawn back to hit him when the African policeman said to me, "please don't do that, man." '

His family were used to him. His wife remained silent throughout the evening except to moo at the children and to say pleadingly to the elder boy, who had been surreptitiously having some of his father's beer with his father's connivance; 'Would you like some lemonade?'

Her son replied: 'No, mam, I don't want anything at all.'

She cuddled her exhausted younger son, and sat with empty eyes. Her husband drank pint after pint, talking continually amid his silent and worn-out family.

He said: 'The trouble is, a lot of kids these days, they like being cosseted, or going to the pictures.'

He got himself yet another drink, this time including his wife, not that he had asked her. She got a drink every now and then, in the even course of his drinking. He now was happy and was telling the bar about the Leopard Men in West Africa.

'That child,' Paul said to me, 'ought to be in bed.'

Twenty minutes later the family left, the elder boy having been bribed with half a dozen bags of crisps. His father, on the way out, put his arm on the boy's shoulder, proudly, protectively and by way of assistance; and from underneath the table where they had sat a great Alsatian got up and followed them out.

Leaving, the Humberman said: 'It's awful in Africa. The white man's burden.'

The old man, who had listened to him throughout, said: 'Yes, it's time. It's the time. It's time to go. They're all Boers now.'

'What a marvellous non-sequitur,' said Paul.

A few more locals came in, I recall: runty, squinty men and youths and coarse laughing witless girls. A place to visit, not to live in, this farthest north-western part of the British mainland; beautiful the land, and the people bred out.

CHAPTER FIVE

Our dispute at Tongue – the Naver valley and the
Clearances – a Highland Gentleman – south-eastwards
to the North Sea – Dunrobin Castle and the Suther-
lands – a mad museum – two boys from Embo – re-
flections at Dornoch – golfers making drunken jokes –
how Paul and I survived each other – a picnic up the
glen – and sense of pilgrimage now at an end – and a
good jaunt we had of it.

We drove around Loch Eriboll after we left Durness, then east
across to the Kyle of Tongue; and at Tongue we had our only
dispute. I wished to push across the road along the top, to
Thurso and to Dunnet Head, arguing that as we had come to
the north, to the northernmost point of the mainland we
should go. I argued this fiercely, the more so since I was
conscious of the weakness of the argument. Paul argued
fiercely, too, wishing to drive down Strath Naver, most celebrated of the valleys of the Highland clearances, and not
caring to go across into Caithness, saying that it was Lowland,
and not Highland and Island. Paul won the argument.

We went east a few miles, then turned south down Strath
Naver, pausing only briefly to look south-westwards through
the Borgie valley to Ben Loyal above Loch Loyal: a mountain
in its solitary splendour; the equal, if not the superior, to my
mind, of the lovely and famous Suilven. About this, too, Paul
and I disputed, he preferring Suilven. It was easy, as we
talked, to see how personality could be ascribed to mountains
and other natural phenomena such as lakes and waterfalls. We
had no difficulty in talking anthropomorphically, and certainly we did not think at the time we talked gibberish,
although we knew we were talking playfully.

Like the Kyle of Durness the Naver valley was tidier,
balmier, more southern than I had anticipated. To look at, it
presented no sense of tragedy or of cruelty. It declined to live

up to its repute. I suppose Paul and I, in our different ways, had hoped there would be stark and bleak evidence of the Sutherland clearances here, in the most celebrated of the cleared places, where Patrick Sellar, as factor of the Countess of Sutherland and as entrepreneur in his own right, had done his worst.

Later I read *The Trial of Patrick Sellar* by Ian Grimble, and recommend it to those who would understand this part of Britain, its emptiness, its blood-curdling beauty. Also, it becomes possible to understand how the enemy of each Scot has usually been another Scot; that it is not only the Highlanders and Islanders and Celts who had the Gaelic disputing with the English-speaking Lowlanders and Norsemen, but that the Highlanders disputed among themselves and the Islanders also, although less so; and that the Highlanders themselves were fighting men, useful as such to their chiefs, but useless when the jurisdiction of the chiefs was ended. Writes Eric Linklater, in his introduction to Grimble's *The Trial of Patrick Sellar*: 'Not all (chiefs) had been wise or humane in the exercise of their hereditary powers, but abolition of them made the whole clan bankrupt; for what the worth of a clansman who was under no compulsion to obey? The value of men declined, the value of money went up. But where was the money to come from? The clan chiefs had no resources but their land, and their land was encumbered by a host of unprofitable tenants. Temptation came from the south of Scotland, where sheep-farmers had discovered the riches inherent in a flock of Cheviots . . . The Highlanders had to go, and go they did; they were, after all, no longer an asset to their chiefs.' Grimble quotes the Sutherland factor Evander MacIver, who came later than the dreadful Sellar and his equally dreadful associate James Lock – the two chief instruments by which the Leveson–Gower family (George Leveson–Gower became the first Duke of Sutherland) impoverished the north in search of wealth from sheep. Evander MacIver wrote, in his *Memoirs of a Highland Gentleman*: 'There is no duty I performed during my services as factor in Sutherland on which I look back with more satisfaction than the time, trouble and care I expended in carrying out the transportation of so many families from the poor position of crofters in a wet climate and a poor soil for cultivation, to the more fertile lands of Canada, Nova Scotia and Australia. The crofter system has not within it the seeds of prosperity or of profit.'

Grimble tells us that when this nauseatingly snobbish Scotsman died and his memoirs were posthumously published, although the Dukes of Teck, Sutherland and Westminster subscribed, in the list of subscribers none was called Mackay: and it was the Mackay country, the Reay forest, in which, at Scourie, he had lived for sixty years.

The Scots have produced some terrible monsters to make their English lords wealthier. The Highland aristocrat, Lord Reay himself, in the end was no better than the rest: for he sold the last hundreds of thousands of acres of the Mackay lands in 1829, to the Countess of Sutherland. The Chief of Mackay, the most Highland of all Highlanders, sold out; and it took a Swiss sociologist to write, in 1837: 'Before reaching such a barbarous resolution it had been necessary for the noblemen to cease utterly to share the views, attachment and sense of decency of his fellow men. It had been necessary not merely for him to believe himself no longer their father or their brother, but even to have ceased to believe himself of the same race. It had been necessary for an ignoble greed to extinguish in him the sense of consanguinity to which their ancestors had trusted when they bequeathed the destiny of their people to his good faith.'

Driving down the valley of the Naver we saw some sporting fishermen, discussing some sporting point. All in all, the beauty of the north-western highlands is a terrible beauty, a terrible beauty that was still-born. It is a beauty enough, when you think about its wasted lands and empty places and vast distances, to crack your guts open.

Paul and I chatted mildly on the road south and east, until we came across the North Sea, and a kind of prosperity, a degree of long-lived, peaceable, not wholly impoverished and improvident settlement.

We came, a day or two later, to look around Dunrobin Castle, the great palace of the Dukes of Sutherland. Dunrobin

Castle is now a school. In the grounds there is a family museum, the like of which neither Paul nor I had seen nor thought ever to see. For instance, there is a framed address: 'To his Grace Cromartie, Duke and Earl of Sutherland: we the undersigned Crofters in the Parish of Lairg desire to express to Your Grace our warmest thanks for, and heartfelt appreciation of Your Grace's generosity in granting us a reduction of fifty per cent on our rents.

'We recall with the sincerest gratitude that this is not the only occasion on which we have experienced Your Grace's kindness.

'Last year it was Your Grace's pleasure to give us a similar reduction. Such benevolent deeds on the part of Your Grace will go a great deal to further strengthen the cordial relations which have hitherto existed between Your Grace and Your Grace's tenants.

'We are exceedingly pleased that this generous and noble act of Your Grace has given an opportunity to all Your Grace's tenants to meet their legal obligations to Your Grace.

'It is our earnest prayer that Your Grace and Her Grace the

Duchess of Sutherland may be long spared in the enjoyment of health and that it will be at all times a pleasure to Your and Her Grace to promote the interests of Your Grace's tenants.'

This document was signed in 1894, long after the clearances. Among the museum's other proud exhibits are vast numbers of slaughtered animals, an antique Baby Vacuum Cleaner, the telephone from Dunrobin Private Railway Station, a case containing the 'Full Dress and Active Service Uniform and Accoutrements of the late Major the Lord Alistair St. C. Leveson-Gower, M.C.', elephants' feet, giraffes' heads and necks, an alligator stuffed in a begging position, several elephants' tails and tusks, tails of white rhinos, sweet little heads of dik diks and other wild animals all laughingly shot by Lord and Lady Stafford and the Duke and Duchess; a greater falcon and countless other lovely birds; the figure-head from the S.Y. *Catania*; giant beetles from Uganda, moths from Singapore, and a fierce tarpon of 95 pounds 'caught and landed in thirty minutes with Rod and Line in the Shah River, Florida, by the Duke of Sutherland, January 1925'; and also a chair made from the bones of a whale which drifted ashore at Kintradwell in 1869.

'They really were a frightful bunch of hooligans,' Paul said to me, or I said to him, as we left the mad museum.

We then drove to Embo, a fishing village, its houses derelict by the sea's edge, and we gave a lift to two boys who wanted to get to Dornoch.

'What do people do in Embo now the fishing has stopped?' I asked.

'Nothing,' one of the boys said.

'Nothing?'

'People do nothing in Embo.'

One of the things they used to do was to die for their country. In these remote empty northern places, where the towns by English standards are villages, despite the dignified scale of their public buildings, you may find, for example, a sign in Dornoch which reads: 'Erected by the Seaforth Highlanders to the undying memory of 8,432 comrades belonging to the ten battalions of the Regiment who have given their lives for

their country in the Great War. Scotland forever.' That is the
kind of notice to make one weep.

There are others which occasion a smile, for they express the
Scottishness of Scotland equally, but less sadly: 'Bus Stance',
for example, 'The Police Office', 'Sheriff-Clerk's Office',
'Procurator-Fiscal's Office'.

At Dornoch I was also fortunate, benefiting from losing the
earlier dispute. Paul, apologetic for having argued successfully
that we should take the dark and dreary road beyond Strath
Naver, said in the car: 'I think you were right,' and I agreed.

A few miles later on he said: 'Well, George, I shall make a
gesture. I have booked a room in the Dornoch Castle Hotel
and one in the annexe, which was all they had left. The
annexe, the lady said over the telephone, is three hundred
yards up the road and not nearly so nice as the Castle. We
would, I think, have had to toss for it. You shall have the
room in the Castle.'

I replied: 'I accept that in the spirit it was offered.'

I thus ended up with a most delightful room overlooking a
neat formal garden with fields beyond to the Dornoch Firth;
and felt this to be a fair exchange for Dunnet Head and an
extra sixty or eighty miles of driving.

Clear differences in kind, from the Highlands and Islands, are
in evidence at Dornoch. A neat, seemly and indeed becoming
township, of unmistakably Scottish character, yet it was also
Lowland, and English in some of its matters and style. It
was not that we had come south but that we had come east and
had left the sighs and vapours of the west lands behind us.

It is the westwardness and not the height or the insularity
or the nor'wardness which is what the Highlands and Islands,
or the Celtic, thing is about. I do not think it may necessarily
be Celtic, or racial, or even very much linguistic. There is
looniness about the west. The west coast has the loony business in America. The dreamers and fools and poets and weak-
lings, the singers and the charmers, look westwards, longingly,
over the hills and the seas where the sun has been, preferring
the evening above the morning, mystery before fact, idleness
or converse rather than the labour of hands and minds, and
suffering in the long run poverty of mind and body (for al-

though poverty is far from coming to America's west coast, I sometimes feel that it is there that the wealth will run out, just as it is there that the brains and the sands may well already be running out).

A different interpretation is also possible. The hotel, like much of Dornoch, is devoted to golf; and to English visitors. They are not necessarily cultivated people. Indeed to many people's taste, mine included, the sportsmen-visitors to the Highlands and Islands are less intolerable than the golfers. Do barbarians keep the east rich, if necessary by letting the blood of the west?

The bar at the Dornoch Castle hotel was uncongenial the night I spent in it. People were laughing noisily.

'Did you hear the one about the back passage?'

'Ssh.'

'Nasty jokes again!'

They all were heaving, wheezing and heaving, wheezing with suppressed giggles.

She said, '. . . And she said, "Kindly put your umbrella down!" '

He said, 'She said what? Surely that was the wrong thing to say!'

'Oh, oh, oh, oh! My God! Ha, ha, ha!'

They are all getting very drunk.

'Stop using those words,' he said.

'I said, "I did. I stopped. I'm only speaking to the god, I said." '

'You can ask a lady but you cannot ask a man,' one of the women said.

A man said, 'You can ask me, I'll say yes, ha, ha, Oh, My God! Ask me, and I'll say yes, ha, ha, ha!'

The barman then told a story. They all laughed. The barman had earlier bought a drink for a local who had strayed into the bar: a mascot figure, possibly one of the hotel servants.

'Ye're bad, bad,' said this local to the barman.

'Come on, John, come on John,' the English couple shouted, persuading the local to tell his own dirty joke.

'No, no, I'll not tell. Ye're bad. No, but he's bad, an ah'll have another wee drop, but he's bad.'

'Just another wee drop for John,' the English golfer laughed.

He then started rattling off a dirty alphabet, '. . . Q is the quickness with which he gets in, . . .' The women and John and the barman applauded and laughed.

'Ah, ah,' said John. 'Now that's very clever. How can you remember all that?'

'I learnt it all at school,' the Englishman said.

'At school, did you?' John said. 'Remembering all that, now that's very clever.'

We now drove south, to Inverness, where we were to stay for the weekend as guests of Hugh and Antonia Fraser at their island house, Eilean Aigas, in the Beauly river; part of the great Lovat estates. The house party was pleasantly mixed: there was tennis, swimming in the steaming hot open-air swimming pool, salmon-fishing for those who fancied it, and church, too, for those who wanted or needed it; ping-pong, too; a great Common Market debate in which most of us, being opposed, felt unable to deal properly with the arguments of the most vehement pro-Marketeer, for he talked with the thickish accents of a European banker.

There was a picnic miles and miles up the glen where, farther on, was the shortest crossing in Scotland from the east to the west. We could almost have walked back to the west and seen the lochs stretching westwards to the Atlantic. It was said that the Lord Lovat of the day was carried down this glen by his retainers, after the 1745.

The glen was almost unpopulated, except for the hydroelectric people. The exhaust fell off the car, I having lent it to Nigel Ryan to drive some of the Papists to their prayers. Hugh Fraser, however, tied it on again. We drove down to Perth, put the car on the train and arrived in London on the following morning.

We had survived each other remarkably well. The arrangements between each other had been formalised on one occasion after church at Tobermory. I said I wanted to go to Staffa.

Paul replied: 'We don't know if we can get there.'

I said, 'Why don't you go and find out.'

Paul said: 'I'm not your bloody secretary.'

I said: 'Yes, you are. I'm your bloody chauffeur.'

Paul said, 'Oh,' and went off to telephone Ulva Ferry. We did not take that ferry, by agreement, it being inconvenient.

Let me repeat the one residue of the trip that will remain. Later in the summer, as I've mentioned, we drove through Europe down to Cavtat, through Carpathia, among the Alps, almost as far south as the Greek Islands; and I suppose the scenery was no less beautiful than Paul and I had seen in the north-west of Scotland; and the historical associations of the places we visited south of Cavtat were certainly greater, if less reverberative.

I do not mind if I never travel through Carpathia again, or drive to Cavtat, or see the golden and the pink crumbling, sinking stones of Venice. But I will hope to see Mull again, and Applecross; and argue the merits of Suilven and Stac Polly; and hope to see Ben Hope and Ben Loyal beyond and travel eastwards again as far as the Kyle of Tongue, and there to look southward and see Ben Loyal alone, standing up in its northernmost plain, from which people have been driven, and have let themselves be driven: the beauty is partly in the historical sadness and the emptiness.

That sense of pilgrimage which I have mentioned has gone. Perhaps it was because of not knowing to what part of Scotland the Scottish part belonged or because the residual sense of part Scottishness was washed away, without much regret. But I may still day-dream – and indeed, I will, as do all Englishmen who have caught the ineradicable infection of the beauty of the Highlands and Islands.

A good jaunt; a jaunt to be recommended and done again: a jaunt that, in the end, proved itself to be no more than that, and no less either.

INDEX

Index

Index